BASIC surveying

Butterworths BASIC Series includes the following titles:

BASIC aerodynamics
BASIC artificial intelligence
BASIC business analysis and operations research
BASIC business systems simulation
BASIC differential equations
BASIC economics
BASIC electrotechnology
BASIC forecasting techniques
BASIC fluid mechanics
BASIC hydraulics
BASIC hydrodynamics
BASIC hydrology
BASIC interactive graphics
BASIC investment appraisal
BASIC materials studies
BASIC matrix methods
BASIC mechanical vibrations
BASIC molecular spectroscopy
BASIC numerical mathematics
BASIC operational amplifiers
BASIC reliability engineering analysis
BASIC soil mechanics
BASIC statistics
BASIC stress analysis
BASIC surveying
BASIC technical systems simulation
BASIC theory of structures
BASIC thermodynamics and heat transfer

BASIC surveying

W M Barnes BSc, FRICS, FInstCES
Senior Lecturer, Royal Military College
of Science (Cranfield), Shrivenham, England

Butterworths
London Boston Singapore Sydney Toronto Wellington

First published, 1988

© **Butterworth & Co. (Publishers) Ltd, 1988**

British Library Cataloguing in Publication Data
Barnes, W.M.
 BASIC surveying
 1. Surveying. Applications of computer systems. Computer systems. Programming languages. Basic language
 I. Title
 526.9′028′55133

 ISBN 0–408–01248–X

Library of Congress Cataloging-in-Publication Data
Barnes, W. M.
 BASIC surveying/W.M. Barnes.
  p. cm.—(Butterworths BASIC series)
 Bibliography: p.
 Includes index.
 ISBN 0–408–01248–X
 1. Surveying—Data processing. 2. BASIC (Computer program language)
 I. Title. II. Series.
 TA549.B36 1988
 526.9′028′55133—dc19   88—14514

Typeset by August Filmsetting, Haydock, St Helens
Printed and bound by Hartnolls Ltd., Bodmin, Cornwall

Preface

Much of the time and effort in surveying is spent on the processing and manipulation (often repetitive in nature) of considerable amounts of numerical data. Prior to the advent of the electronic computer this task was irksome, time consuming and often very costly. The use of the wide range of computing aids now available reduces this labour and the resultant costs enormously.

A knowledge of what the computer can do and, perhaps more important, the ability to make it do it, has become a necessary part of technical and scientific education. Thus, it is essential for the engineer to have more than a superficial acquaintance with computer programming. Although most technically oriented courses now include the fundamentals of programming, not all demonstrate the relevance of programming to specific fields and students often experience difficulty in applying their programming knowledge to the solution of real problems.

This book makes use of BASIC, an extremely powerful computer language that is easy to learn and use. The simplicity and power of programming in BASIC, and the facility it gives for developing programs using simple English statements without the need for special compilation and editing routines have made the language very attractive for use by the non-professional programmer and, more particularly, by students.

BASIC surveying has been written with the student in mind and is directed primarily to first-year students of surveying and to students in allied fields, including civil engineering, geology and geography, who are commencing surveying courses. The book is not intended as an authoritative text on either computing or surveying; many admirable texts in these fields already exist. The aim is to introduce BASIC by applying it to various computational aspects of engineering surveying. It is hoped that the brief description of the language and the short programs given as examples will familiarize the reader with the elements of BASIC and illustrate the relevance of the language to the solution of fundamental surveying problems. It is also hoped that readers will be encouraged to develop their own programs for the solution of more complex problems.

The programs listed in this book were developed on the Tektronix 4051 microcomputer and Epson MX80 terminal printer. REM (remark) statements have been used throughout the programs to remind the reader of the significance of each routine.

Chapter 1 introduces the concept of BASIC programming and discusses the requirements of program structuring. Chapter 2 presents a brief introduction to surveying principles. Chapter 3 summarizes the fundamentals of survey computations. Chapter 4 is concerned with the provision of survey control. Chapter 5 deals with aspects of site surveying. Chapter 6 concentrates on earthwork calculation. Chapter 7 discusses some computational aspects related to setting out operations.

W.M.B.

To 'M'

Acknowledgements

I wish to express my sincere thanks to Mrs J. G. D. Price of the School of Mechanical, Materials and Civil Engineering, Royal Military College of Science (Cranfield) for her considerable assistance in the typing of the manuscript for this book; to Mr N. Rahman of the Department of Surveying, University of the West Indies, St Augustine, Trinidad for his help in the preparation of a number of the diagrams included in the text; to my wife and family for their forbearance and understanding during the book's preparation; and to my publishers for their fortitude and patience.

Contents

PROGRAMS

PROGRAMS

Chapter 1

Programming in BASIC

1.1 Introduction

BASIC (Beginner's All-Purpose Symbolic Instruction Code) was developed at Dartmouth College, USA, in 1964, as a simple, general-purpose programming language. Since then, extremely powerful versions of BASIC have been developed and the language has increased in popularity among scientific programmers to become the main language used with the modern microcomputer. Its time-sharing capabilities enable the computer to be used from different terminals and interaction between the computer and programmers at each terminal is virtually immediate. The computer pauses to display a question or to receive information and reacts instantly to the programmer's reply. Using the interactional mode, therefore, programs may be typed in at the computer directly, errors corrected, and the program run simply and with little loss of time.

1.2 The basics of BASIC

1.2.1 Statements

The BASIC program comprises a series of statements defining the sequence of steps which the computer is required to follow. Each statement has a similar structure:

line number – keyword – body

The line number is unique and determines the order of execution. Line numbers usually increase in steps of 10, as this facilitates the insertion of additional statements should the need arise in the course of program development. The keyword describes the action of the statement, e.g. READ (data), PRINT (the answer). The body provides details in the form of variables, constants, expressions and functions for the keyword.

1.2.2 Mathematical expressions and operators

Mathematical models form the basis of most surveying computations and comprise numeric variables, constants and functions.

In the BASIC statement numeric variables are represented by a single capital letter or a capital letter followed by a digit and must be so designated in the program.

Numerical constants are entered into the program just as one would write the number, e.g. 123.456; −0.789; 246. Scientific or E notation facilitates the use of very large or very small numbers; for example, −123.456 E − 15 equals −123.456 × 10⁻¹⁵.

Different versions of BASIC provide a variety of mathematical functions but the following are common to all. (The argument (X) may be a number, a variable or a mathematical expression.)

General mathematical functions

ABS(X)	returns the absolute value of X
EXP(X)	returns the value of the base e raised to the X power (eˣ)
INT(X)	returns the largest integer without exceeding X
LGT(X)	returns the logarithm of X to the base 10
LOG(X)	returns the logarithms of X to the base e
PI	returns an approximate value of π
RND(X)	returns a random number between 0 and 1
SGN(X)	returns + 1 if X is positive, 0 if X is zero and − 1 if X is negative
SQR(X)	returns the square root of X

Trigonometric functions

SIN(X)	returns the sine of X
COS(X)	returns the cosine of X
TAN(X)	returns the tangent of X
ASN(X)	returns the arc sine of X
ACS(X)	returns the arc cosine of X
ATN(X)	returns the arc tangent of X

For trigonometric functions the argument (X) is usually interpreted as being measured in radians but some BASIC systems provide for the setting of trigonometric units to degrees or grads as alternatives to radians.

Arithmetic operators

Mathematical models also contain arithmetic operators which together with variables and constants form mathematical expressions. The following arithmetic operators are common to all BASIC systems:

($\hat{\,}$) raise to the power, e.g. $A\hat{\,}2$
(*) multiply $A*2$
(/) divide $A/2$
(+) add $A+2$
(−) subtract $A-2$

Longer expressions may be written containing several operators when the order of evaluation is determined by the following hierarchy: $\hat{\,}$ (highest); * or /; + or − (lowest). For example, when evaluating $A+B*C\hat{\,}4$, C is evaluated first, multiplied by B and added to A. Operators of equal priority are evaluated from left to right through the expression. For example $A-B*C+C*D+E$ is evaluated in the order BC, CD, $A-BC$, $A-BC+CD$ and $A-BC+CD+E$. Parentheses may be used to override any of these operators as an expression enclosed by parentheses is evaluated before any expression outside.

When nested sets of parentheses are used, the computer executes the operations indicated in the innermost pair of parentheses first then works its way outwards using successive pairs of parentheses. An odd number of parentheses represents an ambiguity and would be queried as an error. For example, $(A+B)*C\hat{\,}4$ is evaluated as $A+B$, C^4, $(A+B)C^4$; $A+(B*C)\hat{\,}4$ is evaluated as BC, $(BC)^4$, $A+(BC)^4$; $(A+(B*C))\hat{\,}4$ is evaluated as BC, $A+BC$, $(A+BC)^4$.

1.2.3 Subscripted variables

Named exactly as for unsubscripted variables (a letter or a letter and one or two digits), subscripted variables are differentiated by one or two subscripts enclosed in brackets after the name. A singly subscripted variable cannot have the same name as a doubly subscripted variable but both a subscripted and unsubscripted variable of the same name may appear in the same program. For example, A(1), B(10), A1(3,4), B2(99,99) may all be used.

The computer must be informed of the use of all subscripted variables and of the maximum value of each subscript by means of a DIM statement in the form:

line number – DIM variable 1 (integer 1), variable 2 (integer 2) . . . variable n (integer n)

For example: 100 DIM A(5), B(10), A1(4,4), B2(100,100)

DIM statements may be positioned anywhere in the program but must appear before the subscripted variables are first used. They are conventionally grouped at the beginning of the program.

Subscripts need not be constants. Subscripted variables may be used very effectively in conjunction with the FOR–NEXT loops (Section 1.2.6). This is a useful facility as it allows a single variable to have a number of different values during a single program run.

1.2.4 Assignments

As the computer follows the sequence of numbered statements, values are allocated to each of the variables either directly by data input into the program or by generation following an assignment statement.

The LET statement assigns an initial or new value to a variable and takes the general form:

line number – LET – variable or constant = mathematical expression, variable or constant.

For example:

50 LET I1 = B*D^3/12

The word LET is usually optional and is often omitted from the statement. For example, $100 X = X + 1$ assigns a new value to X of $(X + 1)$.

1.2.5 Input

The input of numeric data may be effected in one of two ways.

Before the program is run a READ statement defines the variables to which values contained in an associated DATA statement are to be allocated. The two statements take the following form:

line number – READ – variable 1, variable 2 . . . variable n

For example:

50 READ A, B, C

line number – DATA – value 1, value 2 . . . value n

For example:

60 DATA 120, 200, 350

DATA statements need not necessarily follow directly after the READ statement. They may be inserted anywhere in the program but it is usually more convenient to group them together either at the beginning of the program or at the end. The data for several READ statements can be provided in a single DATA statement having the appropriate number of values as the action of READ statements is to take successive values from a data list.

The INPUT statement provides an alternative form for data entry and is convenient for introducing data during the running of the program. The statement requests data to be entered via the terminal and takes the form:

line number – INPUT – variable 1, variable 2 . . . variable n

For example:

100 INPUT A, B, C

On reaching this statement the computer prompts the user with a ? and accepts values separated by commas which are typed at the keyboard. For example, using the above example ? 5, 10, 15 would allocate the values 5, 10 and 15 to the variables A, B and C respectively.

1.2.6 Program control

Control statements are used to control the route through the program by unconditional or conditional branching, repeating groups of statements or stopping the program.

RUN places the system under program control.

STOP stops program execution.

END ends program execution.

GOTO transfers control unconditionally to a specified statement number. It takes the form:

line number – GOTO – line number

For example, 80 GOTO 100 transfers control directly to statement 100 and 80 GOTO 20 transfers control back to statement 20. By using the GOTO statement a repetition of procedures or looping may be achieved.

IF THEN transfers control conditionally depending on the result of a 'true' or 'false' test and may take one of the following forms:

line number – IF – relational expression – GOTO – line number
line number – IF – relational expression – THEN – line number
line number – IF – relational expression – THEN – statement

The relational expression compares the values of two arithmetic expressions by means of the following relational and conditional operators:

=	is equal to	e.g.	$A = B$
<	is less than		$A < B$
>	is greater than		$A > B$
< =	is less than or equal to		$A < = B$
> =	is greater than or equal to		$A > = B$
< > =	is not equal to		$A < > = B$

If the relational expression is true, control is transferred to the stated line number or the statement immediately following THEN is executed. If the expression is false, control is transferred to the line following the IF statement. For example:

100 IF B = 0 GOTO 200
200 STOP

FOR and NEXT are two statements which are always used together and which enable repeated execution of a group of statements. They take the form:

line number – FOR – variable = value 1 to value 2 STEP value 3
line number – NEXT – variable

For example:

50 FOR I = 1 to 10 STEP 2
100 NEXT I

The FOR and NEXT statements are separated by other statements and the values of the variable (I in the example) is changed successively by the STEP value (2) in the stated range (1 to 10). I acts as a counter recording how many times the group of statements between the FOR and NEXT statements have been repeated. Following the last execution, control moves to the statement immediately after the NEXT statement. Should no STEP value be indicated it is assumed to equal 1.

ON – THEN, ON – GOSUB (mutliple branching) statements take the form:

line number – ON – expression – THEN – line number
line number – ON – expression – GOSUB – line number

On reaching one of the statements the program transfers to the terminating line number if the integer value of the expression equals the terminating line number. For example:

100 ON Z THEN 1000
50 ON Z*2 GOSUB 2000

GOSUB transfers control to the first line statement of a subroutine. The GOSUB statement is of the simple form:

line number GOSUB line number

For example, 100 GOSUB 200 would transfer control from line 100 directly to line 200.

RETURN transfers control from the subroutine to the statement immediately following the GOSUB command. The RETURN statement is simply:

line number RETURN

Subroutines are useful when the same series of statements needs to be accessed several times in the same program or a different program. The use of a subroutine obviates the need for wasteful repetition of the same statements. Subroutines may be placed anywhere in the program but are usually positioned at the end.

Other useful statements

REM is a remark or explanatory statement that the computer ignores in executing the program but which is included in the program listing. The REM statement enables the user to punctuate the program with comments explaining the significance of the various program steps and takes the general form:

line number REM comment

User definable functions may be created using a DEF statement. For example, 100 DEF FNA(X) = 5^X defines the function of A as 5^X, and 110 J = FNA(7) evaluates the function of A(5^X) using 7 for the value of X. The result is assigned to the variable J.

1.2.7 String variables

Non-numeric data can be handled using string variables, a string comprising a series of characters within quotes, e.g. "SURVEY", and a string variable is a letter followed by a $ sign. For example, 100 DIM A$(8), B$(200) dimensions A$ to 8 characters maximum and B$ to 200 characters maximum and 200 LET C$ = "SURVEYING" assigns the string "SURVEYING" to C$.

1.2.8 Output

The PRINT statement is used for outputting data and the results of calculations performed by the program and takes the form:

line number PRINT list

and the list may include variables or expressions, text enclosed in quotes, or a mixture of text and variables. The items in the list are separated by commas or semicolons. A line of print consists of five fields, each of 14 columns, and the action of the separating commas is to move the printing position to the beginning of the next field. Numbers are printed left justified in each field, in decimal notation for numbers 0.01 and 99999, otherwise in E notation. Fields may be left blank by inserting additional commas. Values may be printed out in a compressed form by using semicolons as separators when numbers are printed with only a single space, excluding the sign, between.

Chapter 2

Introduction to surveying

Surveying is the technical term given to the science of measuring and delineating the physical features of the earth and of works executed or proposed upon its surface. It is concerned with the determination and representation of relative spatial positions: the positions of physical and cultural features in topographic surveys; of legal boundaries, real or imaginary, in cadastral surveys; of faults, mineral bearing strata and tunnel centre-lines in mining surveys; of the sun, stars and artificial satellites in geodetic surveys; and of structures and earthworks in engineering surveys.

Surveying is also concerned with the determination of the spatial extent of features, with areas of properties, with volumes of earthworks, with the capacity of dams and the discharge of rivers, and with the size and shape of the earth itself.

The location, form and dimensions of such features are determined by measurement of distance, angles (in both the horizontal and vertical planes) and directions, measurements that are subjected to reduction and adjustment procedures prior to their use as numerical data or as the basis for graphical representation. Photogrammetry, remote sensing, artificial satellite observations, and inertial techniques are among modern systems that, together with the more conventional methods, are used to collect data used in the surveying process. The relatively easy access to electronic computers now facilitates the storage and rigorous processing of large quantities of such data.

2.1 Engineering surveying

A large proportion of all surveys undertaken are concerned with the conception, design and execution of engineering works including site surveys and surveys connected with the positioning and monitoring of structures. Every construction project of any magnitude is based to a large extent upon measurements taken during a survey and developed about lines and points established by the surveyor.

Engineering surveying therefore involves the preparation of

9

topographic maps and site plans, generally at large scale, that are used as a basis for the design of engineering works. The distinctive aspect of engineering surveying is that the positions to be occupied by salient features of new works must then be set out and defined on the ground within specified tolerances.

2.2 Precision of measurements

Physical measurements acquired in the process of surveying are correct only within certain limits because of errors that cannot be totally avoided. The degree of precision of a given measurement depends to a large extent on the methods and instruments employed. It is often desirable that all measurements be made with high precision but, unfortunately, a given increase in precision is usually accompanied by more than a proportional increase in time, effort and cost. It therefore becomes the duty of the surveyor to recognize the requirements and maintain a degree of precision compatible with the purpose of the survey, but not higher. To do this the surveyor must have a thorough knowledge of the sources and type of errors, the effects of errors upon field measurements, the instrument and methods to be employed to keep the magnitude of the errors within allowable limits, and the intended use of the survey data.

2.3 Principles of surveying

Though the advanced stages of geodesy, photogrammetry, error analysis, figural adjustment etc require a sound knowledge of mathematics, the basic principles of surveying are few and relatively straightforward. A knowledge of geometry and trigonometry is essential and, to a lesser degree, of physics, the behaviour of random variables and numerical analysis. Some knowledge of mathematical statistics is invaluable for an understanding of error propagation which, in turn, is needed for the structuring of survey procedures and the selection of equipment. In this respect, a knowledge of differential calculus is also necessary.

Data processing with high-speed electronic computers is so much a part of the recording, reduction, storage and retrieval of survey data that familiarity with such systems is now deemed essential for the modern surveyor.

The measurement operations and computational techniques of surveying are many and various, yet underlying them all are some fundamental principles which provide a unity and a discipline to the subject.

1. 'To work from the whole to the part' is an old surveying maxim which means that no matter what size the survey the main framework should be set out on as large a scale as possible and involve the minimum possible number of measurements thereby reducing the effect of those errors inherent in all measurements. Thus, the accumulation of errors throughout the survey is controlled.
2. It is important to choose the method of survey appropriate to the required result. Every survey is performed for a specific purpose and for each such purpose suitable specifications for the accuracy of the required survey must be devised. Hence, the more refined the techniques and instruments employed, the greater the accuracy obtained. In the interest of speed and economy, the surveyor should work as close as possible to the limits of allowable error.
3. It is important to provide adequate checks. A survey must be designed in such a way that it becomes impossible for errors to pass undetected. This necessitates the inclusion of checks. In fieldwork these take the form of redundancies, i.e. measurements in excess of the minimum numbers necessary to fulfil the geometrical requirements of the survey. There is no such thing as an absolutely exact measurement and the precision (itself a relative term) achieved depends on the instrument used in making the measurement, the care taken, and the ability of the surveyor to guarantee that his work is error free. It is essential that surveying work should be self-checking wherever possible. No confidence may be placed in results that have not been checked. When a computer program known to be operating properly is used for the solution, checking should be concentrated on the input listings to the computer to eliminate tabulation errors and rounding off anomalies in respect of data variables and constants, and to demonstrate the reliability of the results.

2.4 Units of measurement

The operations of surveying entail both angular and linear measurements together with derived area and volume quantities.

2.4.1 Angular units

The fundamental unit of a plane angle is the radian (rad) defined as the angle subtended at the centre of a circle by an arc of circumference equal in length to the radius of the circle. There are thus 2π radians in one complete revolution where $\pi = 3.141\,592\,654\ldots$

The sexagesimal units of angular measure are the degree, minute and second. A plane angle extending completely around a point equals 360 degrees (°); $1° = 60$ minutes ('), and $1' = 60$ seconds ("). We therefore have the following relationship:

$$360° = 2\pi \text{ rad} \quad 1° = 0.017\,453\,283 \text{ rad} \quad 1 \text{ rad} = 57.°295\,779\,5$$

The number of seconds of arc in a radian (206 265") is a frequently-used conversion factor in surveying. It is the factor for converting radians to seconds and it occurs mainly in the determination of small angular corrections by differential methods in which angular quantities are expressed in natural units (radians).

In the centesimal system 400 grades (g) = 360°:

$$1^g = 100 \text{ centesimal minutes } (^c) = 0°54'00''$$
$$1^c = 100 \text{ centesimal seconds } (^{cc}) = 0°00'32.4''$$

Centesimal minutes and seconds can therefore be expressed directly as decimals of a grade, e.g. $49^g\,76^c\,97^{cc} = 49.7697^g$.

2.4.2 Linear units

The international unit of linear measurement is the metre defined in 1983 by the Système International d'Unités (SI) as a length equal to the length of the path travelled by light in vacuum during a time interval of $1/299\,792\,458$ of a second.

The metre is divided into the following units:

1 kilometre (km) = 1000 metres (m)
1 decimetre (dm) = 0.1 m.
1 centimetre (cm) = 0.01 m
1 millimetre (mm) = 0.001 m
1 micrometre (μm) = 0.001 mm = 10^{-6} m
1 nanometre (nm) = 0.001 μm = 10^{-9} m

2.4.3 Units of area

The SI metric units of area are:

1 are (a) = $(10\,\text{m})^2 = 10^2\,\text{m}^2$
1 hectare (ha) = $(100\,\text{m})^2 = 10^4\,\text{m}^2$
1 square kilometre $(\text{km})^2 = (1000\,\text{m})^2 = 10^6\,\text{m}^2$

Chapter 3

Survey computations

3.1 Accuracy and precision

The term *accuracy* refers to the closeness between measurements and their expected values (or their true values). The further a measurement is from its expected value, the less accurate it is.

Precision, on the other hand, pertains to the closeness to one or another of a set of repeated observations of a random variable. Thus, if such observations are closely clustered together, then the observations may be precise but not accurate if they are closely grouped together about a value that differs from the expectation (or true value) by a significant amount. Conversely, observations may be accurate but not precise if they are well distributed about the expected value but are significantly different from each other. Observations will be both precise and accurate if they are closely grouped around the expected value.

3.2 Errors and adjustments

Variability in repeated measurements (under similar measuring conditions) is an inherent characteristic of observations. Observations and measurements are numerical values for random variables which are subject to statistical fluctuations. The term *error* can be considered as referring to the difference between a given measurement and a true value of the measured quantity. As the latter is almost never known the exact magnitude of the error is therefore unknown.

All survey operations are subject to error and a knowledge of error characteristics, their magnitude and their behaviour is essential in order to be able to assess whether observations conform to the standards of accuracy appropriate to the particular technique being used. Such knowledge is also an important factor in deciding on the methods to be used in any project, especially when the accuracy required has to be considered in relation to time and expense.

Assuming the observations are free from gross error (mistakes and blunders) and that systematic errors have been corrected, the observations must be adjusted to give their *most probable values*. In surveys of limited scope it is permissible to use simple methods for adjusting accidental errors, methods that are not necessarily based on the laws of probability. The use of such methods may save considerable time and they can be easily learned and applied.

3.3 Computing procedures

Calculations of one kind or another form a very large part of the work of surveying and the ability to compute with speed and accuracy is an important qualification of the surveyor.

Several computational steps are fundamental to all surveys and the frequency with which they are required and used is reflected by the fact that many pocket calculators now incorporate in their design the facility for 'one step' solution of some basic calculations. For instance the conversion of angular measure from sexagesimal to decimal or radians is now almost universally available as is the conversion from polar to rectangular coordinates and vice versa. However, these facilities are not always available in micro or mainframe computers and conversion subroutines must therefore be introduced in programs that require the frequent use of such conversions.

Being fundamental computations, they also serve to introduce both fundamental theory and simple examples of the use of BASIC programming.

3.3.1 Plane rectangular coordinates

Let it be assumed that an area under survey is not too extensive (say less than 1 km in any direction) and that the effect of earth curvature is negligible. A system of plane rectangular coordinates may then be used to relate points forming a control framework. Such a system could be allocated a false origin, with coordinate values (0,0) say, and preferably chosen to the extreme south and west so as to conveniently produce positive coordinates for all points in the area under survey.

The scale along both axes is the same and coordinates are reckoned positive from west to east and from south to north. Figure 3.1 shows the relative positions of two control points, A and B together with their rectangular coordinates (E_A, N_A) and (E_B, N_B) respectively, and the partial rectangular coordinates ΔE_{AB} and ΔN_{AB}.

The *bearing* of one point from another is defined as the clockwise horizontal angle that the line joining the two points makes with some fixed direction. In plane surveying this direction is usually *grid north* defined as the direction parallel to the northward pointing axis of the

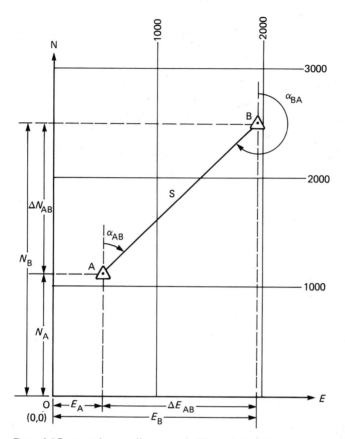

Figure 3.1 Rectangular coordinate system. The 'polar' and the 'join'

plane rectangular coordinate system. α_{AB} is therefore the bearing of the line A to B. The reverse bearing B to A (α_{BA}) is then simply $\alpha_{AB} \pm 180°$, i.e. the reverse bearing equals the forward bearing $\pm 180°$.

The polar coordinates of point B with respect to point A therefore comprise the bearing α_{AB} and the horizontal distance S between the two points.

The relationship between the polar coordinates and the rectangular coordinates of A and B are therefore:

$$E_B - E_A = \Delta E_{AB} = S \sin \alpha_{AB} \quad \text{and} \quad E_B = E_A + S \sin \alpha_{AB} \left.\right\}$$
$$N_B - N_A = \Delta N_{AB} = S \cos \alpha_{AB} \quad \text{and} \quad N_B = N_A + S \cos \alpha_{AB} \left.\right\} \quad (3.1)$$

Example 3.1

Given the following data determine the coordinates of point B (the 'polar'):

Coordinates

	E (m)	N
A	450.16	1126.79

Grid bearing AB $= 47°13'24''$ Distance AB $= 2032.77$ m

Solution:

$E_B = E_A + S \sin \alpha_{AB} = 450.16 + 1492.066 = 1942.226$
$N_B = N_A + S \cos \alpha_{AB} = 1126.79 + 1380.540 = 2507.330$

Program 3.1 Conversion from polar to rectangular coordinates

```
EASTINGS OF A=450.16
NORTHINGS OF A=1126.79
BEARING A TO B=47.2233333333
DISTANCE A TO B=2032.77
EASTINGS OF B=1942.22639646
NORTHINGS OF B=2507.33038023
```

```
100 PRINT "PROGRAM 3.1"
110 REM THIS PROGRAM CALCULATES THE RECTANGULAR COORDINATES OF POINT B
120 REM GIVEN THE COORDINATES OF POINT A AND THE BEARING AND DISTANCE
130 REM FROM A TO B
140 INIT
150 REM RETURNS THE SYSTEM ENVIRONMENTAL PARAMETERS TO A KNOWN STATE
160 SET DEGREES
170 REM SETS THE TRIGONOMETRIC UNITS TO DEGREES
180 READ D,M,S,E,N,L
190 DATA 47,13,24,450.16,1126.79,2032.77
200 B=D+M/60+S/3600
210 REM THE BEARING A TO B IS CONVERTED FROM SEXAGESIMAL TO DECIMAL AND
220 REM THE COORDINATES OF A AND THE DISTANCE FROM A TO B ARE ALLOCATED
230 REM TO THE VARIABLES IN 180
240 X=E+L*SIN(B)
250 Y=N+L*COS(B)
260 PRINT @4:"EASTINGS OF A=";E
270 PRINT @4:"NORTHINGS OF A=";N
280 PRINT @4:"BEARING A TO B=";B
290 PRINT @4:"DISTANCE A TO B=";L
300 PRINT @4:"EASTINGS OF B=";X
310 PRINT @4:"NORTHINGS OF B=";Y
320 STOP
330 END
```

For the inverse case where rectangular coordinates of points A and B are given and the bearing and distance between the two points are required then:

$$\left. \begin{array}{l} \Delta E_{AB} = E_A - E_B \quad \text{and} \quad \Delta N_{AB} = N_B - N_A \\ \tan \alpha_{AB} = \dfrac{\Delta E_{AB}}{\Delta N_{AB}} \quad \text{or} \quad \cot \alpha_{AB} = \dfrac{\Delta N_{AB}}{\Delta E_{AB}} \end{array} \right\} \qquad (3.2a)$$

and

$$S = (\Delta E_{AB}^2 + \Delta N_{AB}^2)^{1/2} = \Delta E_{AB} \cosec \alpha_{AB} = \Delta N_{AB} \sec \alpha_{AB} \qquad (3.2b)$$

Which of the two alternative bearing formulae is to be used depends on the magnitude of α and its $\tan/\cot$ function. The function whose absolute value is less than unity is chosen.

Example 3.2
The rectangular coordinates of A and B are as follows:

	E	(m)	N
A	450.16		1126.79
B	1942.23		2507.33

Calculate the polar coordinates (the 'join') between A and B.

Solution:

$$\Delta E_{AB} = E_B - E_A = 1492.07$$
$$\Delta N_{AB} = N_B - N_A = 1380.54$$

$$\cot \alpha_{AB} = \frac{\Delta N_{AB}}{\Delta E_{AB}} = \frac{1380.54}{1492.07} = 0.925\,251$$

and:

$$\alpha_{AB} = 47°13'24''$$
$$S = (\Delta E_{AB}^2 + \Delta N_{AB}^2)^{1/2} = [(1492.07)^2 + (1380.54)^2]^{1/2} = 2032.77$$

Program 3.2 Conversion from rectangular to polar coordinates

```
EASTINGS A=450.16
NORTHINGS A=1126.79
EASTINGS B=1942.23
NORTHINGS B=2507.33
BEARING A TO B=47.2234101799
DISTANCE A TO B=2032.77238679


100 PRINT "PROGRAM 3.2"
110 REM THIS PROGRAM DETERMINES THE BEARING AND DISTANCE BETWEEN TWO
120 REM POINTS A AND B GIVEN THEIR RECTANGULAR COORDINATES
130 INIT
140 SET DEGREES
150 DIM E(2),N(2)
160 REM DEFINES THE UPPER BOUNDS OF THE DIMENSIONS OF ARRAYS E AND N
170 PRINT "ENTER EASTINGS (1)=";
180 INPUT E(1)
190 PRINT "ENTER NORTHINGS (1)=";
200 INPUT N(1)
210 PRINT "ENTER EASTINGS (2)=";
220 INPUT E(2)
230 PRINT "ENTER NORTHINGS (2)=";
240 INPUT N(2)
250 REM VALUES ALLOCATED TO THE VARIABLES E(1),E(2),N(1),N(2)
```

```
260 A=E(2)-E(1)
270 B=N(2)-N(1)
280 C=ATN(A/B)
290 IF C>0 THEN 350
300 IF A<0 THEN 330
310 C=180+C
320 GO TO 370
330 C=360+C
340 GO TO 370
350 IF A>0 THEN 370
360 C=180+C
370 D=SQR(A^2+B^2)
380 PRINT @4:"EASTINGS A=";E(1)
390 PRINT @4:"NORTHINGS A=";N(1)
400 PRINT @4:"EASTINGS B=";E(2)
410 PRINT @4:"NORTHINGS B=";N(2)
420 PRINT @4:"BEARING A TO B=";C
430 PRINT @4:"DISTANCE A TO B=";D
440 STOP
450 END
```

3.3.2 The solution of triangles

A common requirement in surveying practice is the calculation of the rectangular coordinates of a point using angular observations from two other points, the coordinates of which are known.

Figure 3.2 illustrates the situation where the coordinates of A and B are given and the angles β_1 and β_2 have been observed. It is required to determine the coordinates of point C.

The computational sequence is as follows:

Alternative 1

1. Calculate α_{AB}, α_{BA} and S from Equations (3.2a) and (3.2b).
2. Deduce the angle β_3 ($= 180° - (\beta_1 + \beta_2)$).
3. Determine the lengths AC and BC using the sine rule of plane trigonometry:

$$BC = a = \frac{S}{\sin \beta_3} \sin \beta_1, \quad \text{and} \quad AC = b = \frac{S}{\sin \beta_3} \sin \beta_2$$

4. Deduce α_{BC} ($= \alpha_{BA} - \beta_2$) and α_{AC} ($= \alpha_{AB} + \beta_1$) taking care with the signs of the angles β_1 and β_2.
5. Calculate the coordinates of C from A and B using Equation (3.1) providing a check on the computation.

Example 3.3

Using the coordinates of A and B given in Example 3.2 determine the coordinates of C (situated to the right of AB) using the following data: angle BAC $= \beta_1 = 65°10'22''$, angle ABC $= \beta_2 = 42°51'43''$.

Solution:

1. From Example 3.2, $\alpha_{AB} = 47°13'24''$
$$\alpha_{BA} = 180 \pm \alpha_{AB} = 227°13'24''$$

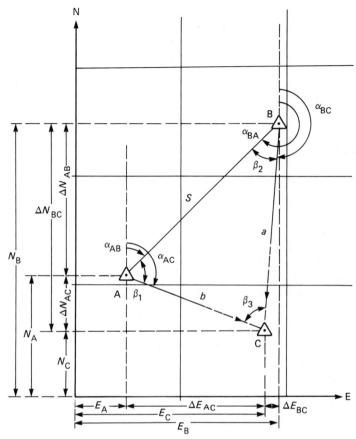

Figure 3.2 Intersection

2. Angle $ACB = \beta_3 = 180 - (\beta_1 + \beta_2) = 71°57'55''$
3. From Example 3.2, $S = 2032.77$ m

Then $BC = \dfrac{S}{\sin \beta_3} \sin \beta_1 = \dfrac{2032.77}{0.950\,869} \times 0.907\,578 = 1940.22$ m

and $AC = \dfrac{S}{\sin \beta_3} \sin \beta_2 = \dfrac{2032.77}{0.950\,869} \times 0.680\,234 = 1454.21$ m

4. $\alpha_{AC} = \alpha_{AB} + \beta_1 = 47°13'24'' + 65°10'22'' = 112°23'46''$
   $\alpha_{BC} = \alpha_{BA} - \beta_2 = 227°13'24'' - 42°51'43'' = 184°21'41''$
5. $E_C = E_A + AC \sin \alpha_{AC} = 450.16 + 1344.52 = 1794.68$
   $= E_B + BC \sin \alpha_{BC} = 1942.23 - 147.55 = 1794.68$ (check)

$$N_C = N_A + AC \cos \alpha_{AC} = 1126.79 - 554.07 = 572.72$$
$$= N_B + BC \cos \alpha_{BC} = 2507.33 - 1934.60 = 572.73 \text{ (check)}$$

Program 3.3 Triangle solution using base angles (1)

```
EASTINGS A=450.16
NORTHINGS A=1126.79
EASTINGS B=1942.23
NORTHINGS B=2507.33
EASTINGS C1=1794.67893096
NORTHINGS C1=572.723935088
EASTINGS C2=1794.67893096
NORTHINGS C2=572.723935088
```

```
100 PRINT "PROGRAM 3.3"
110 REM THIS PROGRAM CALCULATES THE COORDINATES OF POINT C USING THE
120 REM COORDINATES OF POINTS A AND B AND THE ANGLES A AND B
130 INIT
140 SET DEGREES
150 DIM E(4),N(4),D(3),M(3),S(3),A(3),L(3),C(3)
160 PRINT "ENTER EASTINGS, NORTHINGS OF A=";
170 INPUT E(1),N(1)
180 PRINT "ENTER EASTINGS, NORTHINGS OF B=";
190 INPUT E(2),N(2)
200 X=E(1)-E(2)
210 Y=N(1)-N(2)
220 C(1)=ATN(X/Y)
230 L(1)=SQR(X^2+Y^2)
240 REM STEPS 150 TO 220 DERIVE JOIN A TO B (PROG. 3.2)
250 PRINT "ENTER DEGREES,MINS,SECS OF ANGLE A=";
260 INPUT D(1),M(1),S(1)
270 A(1)=D(1)+M(1)/60+S(1)/3600
280 PRINT "ANGLE A=";A(1)
290 PRINT "ENTER DEGREES, MINS, SECS OF ANGLE B=";
300 INPUT D(2),M(2),S(2)
310 A(2)=D(2)+M(2)/60+S(2)/3600
320 PRINT "ANGLE B=";A(2)
330 REM ANGLES A AND B HAVE BEEN CONVERTED TO DECIMAL DEGREES
340 A(3)=180-(A(1)+A(2))
350 PRINT " ANGLE C=";A(3)
360 L(2)=L(1)/SIN(A(3))*SIN(A(1))
370 L(3)=L(1)/SIN(A(3))*SIN(A(2))
380 REM LENGTHS OF SIDES AC AND BC DETERMINED
390 PRINT "BC=";L(2)
400 PRINT "AC=";L(3)
410 C(2)=C(1)+180-A(2)
420 IF C(2)<360 THEN 460
430 C(2)=C(2)-360
440 PRINT "BEARING B TO C=";C(2)
450 REM FORWARD BEARING B TO C DETERMINED
460 E(3)=E(2)+L(2)*SIN(C(2))
470 N(3)=N(2)+L(2)*COS(C(2))
480 REM COORDINATES OF C DERIVED FROM B
490 PRINT "BEARING B TO C=";C(2)
500 C(3)=C(1)+A(1)
510 IF C(3)<360 THEN 550
520 C(3)=C(3)-360
530 PRINT "BEARING A TO C=";C(3)
540 REM FORWARD BEARING A TO C DETERMINED
550 E(4)=E(1)+L(3)*SIN(C(3))
560 N(4)=N(1)+L(3)*COS(C(3))
570 PRINT @4:"EASTINGS A=";E(1)
580 PRINT @4:"NORTHINGS A=";N(1)
590 PRINT @4:"EASTINGS B=";E(2)
600 PRINT @4:"NORTHINGS B=";N(2)
610 PRINT @4:"EASTINGS C1=";E(3)
620 PRINT @4:"NORTHINGS C1=";N(3)
```

```
630 PRINT @4:"EASTINGS C2=";E(4)
640 PRINT @4:"NORTHINGS C2=";N(4)
650 STOP
660 END
```

Alternative 2
If, as is the case in the example, the unknown point lies to the right of
AB, calculate the coordinates of C directly using the observed angles.

$$E_C = \frac{E_A \cot \beta_2 + E_B \cot \beta_1 - N_A + N_B}{\cot \beta_1 + \cot \beta_2} \qquad (3.3a)$$

and:

$$N_C = \frac{N_A \cot \beta_2 + N_B \cot \beta_1 + E_A - E_B}{\cot \beta_1 + \cot \beta_2} \qquad (3.3b)$$

If C lies to the left of AB then

$$E_C = \frac{E_B \cot \beta_1 + E_A \cot \beta_2 - N_B + N_A}{\cot \beta_1 + \cot \beta_2} \qquad (3.4a)$$

and:

$$N_C = \frac{N_B \cot \beta_1 + N_A \cot \beta_2 + E_B - E_A}{\cot \beta_1 + \cot \beta_2} \qquad (3.4b)$$

Example 3.4
Using the data of Example 3.3, calculate the coordinates of C making
direct use of the observed angles at A and B.

Solution:
From Equation (3.3a):

$$E_C = \frac{(450.16 \times 1.077\,563) + (1942.23 \times 0.462\,642) - 1126.79 + 2507.33}{1.540\,204}$$

$$= 1794.68$$

From Equation (3.3b):

$$N_C = \frac{(1126.79 \times 1.077\,563) + (2507.33 \times 0.462\,642) + 450.16 - 1942.23}{1.540\,204}$$

$$= 572.73$$

Program 3.4 Triangle solution using base angles (2)

```
EASTINGS OF A=450.16
NORTHINGS OF A=1126.79
EASTINGS OF B=1942.23
NORTHINGS OF B=2507.33
ANGLE A=65.1727777778
ANGLE B=42.8619444444
```

```
EASTINGS OF C=1794.67893096
NORTHINGS OF C=572.723935088


100 PRINT "PROGRAM 3.4"
110 REM THIS PROGRAM CALCULATES THE COORDINATES OF POINT C USING THE
120 REM COORDINATES OF A AND B AND THE ANGLES AT A AND B. THE POINT C
130 REM LIES TO THE RIGHT OF AB
140 INIT
150 SET DEGREES
160 PRINT "ENTER EASTINGS, NORTHINGS OF A=";
170 INPUT E1,N1
180 PRINT "ENTER EASTINGS, NORTHINGS OF B=";
190 INPUT E2,N2
200 PRINT "ENTER DEGREES, MINS, SECS OF ANGLE A=";
210 INPUT D1,M1,S1
220 A1=D1+M1/60+S1/3600
230 PRINT "ENTER DEGREES, MINS, SECS OF ANGLE B=";
240 INPUT D2,M2,S2
250 A2=D2+M2/60+S2/3600
260 E3=(E1*1/TAN(A2)+E2*1/TAN(A1)-N1+N2)/(1/TAN(A1)+1/TAN(A2))
270 N3=(N1*1/TAN(A2)+N2*1/TAN(A1)+E1-E2)/(1/TAN(A1)+1/TAN(A2))
280 PRINT @4:"EASTINGS OF A=";E1
290 PRINT @4:"NORTHINGS OF A=";N1
300 PRINT @4:"EASTINGS OF B=";E2
310 PRINT @4:"NORTHINGS OF B=";N2
320 PRINT @4:"ANGLE A=";A1
330 PRINT @4:"ANGLE B=";A2
340 PRINT @4:"EASTINGS OF C=";E3
350 PRINT @4:"NORTHINGS OF C=";N3
360 STOP
370 END
```

Alternative 3

Calculate the coordinates of C directly using reduced bearings.

1. Calculate α_{AB} and α_{BA} from Equations (3.2a) and (3.2b).
2. Deduce α_{BC} and α_{AC} applying β_2 and β_1 to α_{BA} and α_{AB} respectively. Then:

$$E_C = \frac{E_A \cot \alpha_{AC} - E_B \cot \alpha_{BC} - N_A + N_B}{\cot \alpha_{AC} - \cot \alpha_{BC}} \qquad (3.5a)$$

and:

$$N_C = N_A + (E_C - E_A) \cot \alpha_{AC} = N_B + (E_C - E_B) \cot \alpha_{BC} \qquad (3.5b)$$

or:

$$N_C = \frac{N_A \tan \alpha_{AC} - N_B \tan \alpha_{BC} - E_A + E_B}{\tan \alpha_{AC} - \tan \alpha_{BC}} \qquad (3.6a)$$

and:

$$E_C = E_A + (N_C - N_A) \tan \alpha_{AC} = E_B + (N_C - N_B) \tan \alpha_{BC} \qquad (3.6b)$$

The choice between each of the above two pairs of equations depends on the absolute value of the terms $(\cot \alpha_{AC} - \cot \alpha_{BC})$ and

$(\tan\alpha_{AC} - \tan\alpha_{BC})$. That pair of equations in which this term approaches closest to unity should be used.

Example 3.5

Again using the data of Example 3.3 calculate the coordinates of C making direct use of the reduced bearings of the lines AC and BC.

Solution:

From Example 3.3:

$\alpha_{AC} = 112°23'46''$ and $\alpha_{BC} = 184°21'41''$

Then using Equation (3.6*a*):

$$N_C = \frac{(1126.79 \times -2.426\,649) - (2507.33 \times 0.076\,268) - 450.16 + 1942.23}{-2.502\,917}$$

$$= 572.73$$

and using Equation (3.6*b*):

$$E_C = 450.16 + [(572.73 - 1126.79) \times (-2.426\,649)] = 1794.68$$
$$= 1942.23 + [(572.73 - 2507.33) \times 0.076\,268)] = 1794.68$$

Program 3.5 Triangle solution using base bearings

```
EASTINGS OF A=450.16
NORTHINGS OF A=1126.79
EASTINGS OF B=1942.23
NORTHINGS OF B=2507.33
BEARING AC=112.396111111
BEARING BC=184.361388889
EASTINGS OF C=1794.68161726
NORTHINGS OF C=572.724937628
```

```
100 PRINT "PROGRAM 3.5"
110 REM THIS PROGRAM CALCULATES THE COORDINATES OF POINT C USING THE
120 REM COORDINATES OF A AND B AND THE BEARINGS FROM A AND B TO C
130 INIT
140 SET DEGREES
150 PRINT "ENTER EASTINGS, NORTHINGS OF A=";
160 INPUT E1,N1
170 PRINT "ENTER EASTINGS, NORTHINGS OF B=";
180 INPUT E2,N2
190 PRINT "ENTER DEGREES,MINS,SECS OF BEARING FROM A=";
200 INPUT D1,M1,S1
210 PRINT "ENTER DEGREES,MINS,SECS OF BEARING FROM B=";
220 INPUT D2,M2,S2
230 B1=D1+M1/60+S1/3600
240 B2=D2+M2/60+S2/3600
250 P=1/TAN(B1)-TAN(B2)
260 Q=TAN(B1)-TAN(B2)
270 IF P-1>Q-1 THEN 310
280 E3=(E1*(1/TAN(B1))-E2*(1/TAN(B2))-N1+N2)/P
290 N3=N1+(E3-E1)*(1/TAN(B1))
300 GO TO 330
310 N3=(N1*TAN(B1)-N2*TAN(B2)-E1+E2)/Q
320 E3=E1+(N3-N1)*TAN(B1)
330 PRINT @4:"EASTINGS OF A=";E1
340 PRINT @4:"NORTHINGS OF A=";N1
```

```
350 PRINT @4:"EASTINGS OF B=";E2
360 PRINT @4:"NORTHINGS OF B=";N2
370 PRINT @4:"BEARING AC=";B1
380 PRINT @4:"BEARING BC=";B2
390 PRINT @4:"EASTINGS OF C=";E3
400 PRINT @4:"NORTHINGS OF C=";N3
410 STOP
420 END
```

3.3.3 *Transformation of coordinates*

Engineering surveys are often undertaken in isolated localities where, for convenience, coordinates of control points are based on a local system with an arbitrary origin and orientation. Should it become necessary to express such coordinates in terms of another system (the national control framework, for example) a coordinate transformation will be required. Figure 3.3 shows two points A and B the coordinates of which are known in terms of each system (a minimum of two points is required for the transformation).

To transform the first system (S_1) with origin of coordinates at O to the second (S_2) with origin at O' requires:

1. a translation of S_1 in eastings and northings by amounts equal to the difference in eastings and the difference in northings between the two origins.
2. a rotation of the whole system about the origin by an amount θ equal to the difference in orientation of the axes of the two systems.
3. a small change of scale which could arise from residual errors occurring in the coordinates of either system.

The general form of such a transformation is as follows:

$$E = PE' + QN' + R$$

and:

$$N = PN' - QE' + S$$

where E and N are the transformed coordinates and E' and N' are the coordinates being transformed.

For the coefficients:

$$P = \frac{(E'_B - E'_A)(E_B - E_A) + (N'_B - N'_A)(N_B - N_A)}{(E'_B - E'_A)^2 + (N'_B - N'_A)^2}$$

$$Q = \frac{(N'_B - N'_A)(E_B - E_A) - (E'_B - E'_A)(N_B - N_A)}{(E'_B - E'_A)^2 + (N'_B - N'_A)^2}$$

$$R = E_A - PE'_A - QN'_A = E_B - PE'_B - QN'_B$$
$$S = N_A - PN'_A + QE'_A = N_B - PN'_B + QE'_B$$

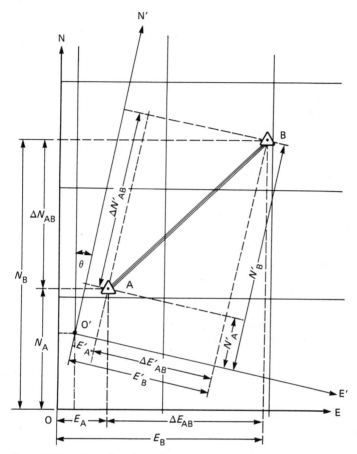

Figure 3.3 Transformation of coordinates

Example 3.6

The coordinates of two points A and B in respect of a local system (E', N') and a national system (E, N) are as follows:

	E' (m)	N'	E (m)	N
A	450.16	1126.79	400 106.53	642 117.86
B	1942.23	2507.33	401 601.71	643 493.23
C	1794.73	572.73		

Determine the transformation parameters to convert coordinates from the local system to the national system. Using the computed parameters derive national system coordinates for C.

Solution:

$$E'_B - E'_A = 1492.07 \quad N'_B - N'_A = 1380.54$$
$$E_B - E_A = 1495.18 \quad N_B - N_A = 1375.37$$
$$(E'_B - E'_A)^2 + (N'_B - N'_A)^2 = 4.132\,163\,576 \times 10^6 = k$$
$$P = [(1492.07 \times 1495.18) + (1380.54 \times 1375.37)]/k$$
$$\quad = 0.999\,395\,703$$
$$Q = [(1380.54 \times 1495.18) - (1492.07 \times 1375.37)]/k$$
$$\quad = 0.002\,905\,858$$

i.e.

$$\text{swing} = \theta = \arctan(Q/P) = 0°10'00$$

and:

$$\text{scale factor} = P/\cos\theta = 0.999\,399\,927$$

$$R = 400\,106.53 - (0.999\,395\,703 \times 450.16)$$
$$\quad - (0.002\,905\,858 \times 1126.79)$$
$$\quad = 399\,653.3677$$

and

$$R = 401\,601.71 - (0.999\,395\,703 \times 1942.23)$$
$$\quad - (0.002\,905\,858 \times 2507.33)$$
$$\quad = 399\,653.3677$$

$$S = 642\,117.86 - (0.999\,395\,703 \times 1126.79)$$
$$\quad + (0.002\,905\,858 \times 450.16)$$
$$\quad = 640\,993.059$$

and

$$S = 643\,493.23 - (0.999\,395\,703 \times 2507.33)$$
$$\quad + (0.002\,905\,858 \times 1942.23)$$
$$\quad = 640\,993.059$$

Then

$$E_C = (0.999\,395\,703 \times 1794.73) + (0.002\,905\,858 \times 572.73)$$
$$\quad + 399\,653.3677$$
$$\quad = 401\,448.68$$

and

$$N_C = (0.999\,395\,703 \times 572.73) - (0.002\,905\,858 \times 1794.73)$$
$$\quad + 640\,993.059$$
$$\quad = 641\,560.23$$

Program 3.6 Two point transformation of coordinates

```
EASTINGS OF A(SYSTEM1)=450.16
EASTINGS OF A(SYSTEM2)=400106.53
NORTHINGS OF A(SYSTEM1)=1126.79
NORTHINGS OF A(SYSTEM2)=642117.86
EASTINGS OF B(SYSTEM1)=1942.23
EASTINGS OF B(SYSTEM2)=401601.71
NORTHINGS OF B(SYSTEM1)=2507.33
NORTHINGS OF B(SYSTEM2)=643493.23
P=0.999395702988
Q=0.00290585817364
ROTATION REQUIRED=0 9 59
SCALE FACTOR REQUIRED=0.999399927537

EASTINGS OF C(SYSTEM 1)=1794.73
NORTHINGS OF C(SYSTEM 1)=572.73

EASTINGS OF C(SYSTEM2)=401448.677461
NORTHINGS OF C(SYSTEM2)=641560.227687
```

```
100 PRINT "PROGRAM 3.6"
110 REM THIS PROGRAM DETERMINES THE PARAMETERS NECESSARY TO TRANSFORM
120 REM COORDINATE SYSTEM 1 (E,N) TO COORDINATE SYSTEM 2 (X,Y)
130 REM WHICH HAS A DIFFERENT ORIGIN, ORIENTATION AND SCALE. THE
140 REM PARAMETERS ARE THEN USED TO TRANSFORM THE SYSTEM 1 COORDINATES
150 REM OF POINT C PROVIDING COORDINATES IN TERMS OF SYSTEM 2
160 INIT
170 SET DEGREES
180 READ E1,N1,E2,N2,X1,X2,Y1,Y2
190 DATA 450.16,1126.79,1942.23,2507.33,400106.53,401601.71,642117.86
200 DATA 643493.23
210 K=(E2-E1)^2+(N2-N1)^2
220 P=((E2-E1)*(X2-X1)+(N2-N1)*(Y2-Y1))/K
230 Q=((N2-N1)*(X2-X1)-(E2-E1)*(Y2-Y1))/K
240 R=X1-P*E1-Q*N1
250 T=Y1-P*N1+Q*E1
260 V=ATN(Q/P)
270 REM V EQUALS THE REQUIRED ROTATION
280 D=INT(V)
290 M=(V-INT(V))*60
300 S=(M-INT(M))*60
310 F=P/COS(V)
320 REM F EQUALS THE REQUIRED SCALE FACTOR
330 PRINT @4:"EASTINGS OF A(SYSTEM1)=";E1
340 PRINT @4:"EASTINGS OF A(SYSTEM2)=";X1
350 PRINT @4:"NORTHINGS OF A(SYSTEM1)=";N1
360 PRINT @4:"NORTHINGS OF A(SYSTEM2)=";Y1
370 PRINT @4:"EASTINGS OF B(SYSTEM1)=";E2
380 PRINT @4:"EASTINGS OF B(SYSTEM2)=";X2
390 PRINT @4:"NORTHINGS OF B(SYSTEM1)=";N2
400 PRINT @4:"NORTHINGS OF B(SYSTEM2)=";Y2
410 PRINT @4:"P=";P
420 PRINT @4:"Q=";Q
430 PRINT @4:"ROTATION REQUIRED=";D;INT(M);INT(S)
440 PRINT @4:"SCALE FACTOR REQUIRED=";F
450 PRINT @4:
460 PRINT "ENTER EASTINGS,NORTHINGS OF C (SYSTEM1)=";
470 INPUT E3,N3
480 X3=E3*P+N3*Q+R
490 Y3=N3*P-E3*Q+T
500 PRINT @4:"EASTINGS OF C(SYSTEM 1)=";E3
510 PRINT @4:"NORTHINGS OF C(SYSTEM 1)=";N3
520 PRINT @4:
530 PRINT @4:"EASTINGS OF C(SYSTEM2)=";X3
540 PRINT @4:"NORTHINGS OF C(SYSTEM2)=";Y3
550 STOP
560 END
```

Chapter 4

Control surveys

Control surveys provide the framework of rigidly fixed points (in terms of coordinates and/or heights) on which site surveys or setting out procedures may be based. The positions of control points are generally determined using triangulation, trilateration or traversing techniques though the use of inertial systems and artificial satellites for position fixing is becoming increasingly more common.

4.1 Triangulation

Prior to the introduction of long-range electromagnetic distance measuring instruments (EDM), triangulation was the sole means of providing control over extensive areas. By measuring the length of one side (a base) and the angles of a triangulation network the lengths of the remaining sides may be determined by trigonometric computation and hence the relative positions of all points included in the network. Ideally, once a triangulation system has been adjusted and coordinates for the triangulation points determined, they themselves provide a basis for subsequent control work.

Engineering surveys are often based on such control but it occasionally happens that construction projects are proposed in localities in which existing control is either too sparse, too distant, or of an order of accuracy inferior to that required by the project specification. In such circumstances it may be necessary to extend the existing triangulation network or to provide a new local system based on an arbitrary origin. Extension obviates the need to provide a measured base as the length and bearing of at least one side of the existing network may be derived from datum coordinates. On the other hand, the provision of a local system will require a measured base and possibly an azimuth determination if sensible orientation of the network is required. The base length would be determined from EDM or catenary tape measurements and the azimuth derived from astronomical or, possibly, gyrotheodolite observations.

4.1.1 Base length reductions

To reduce a measured distance to the horizontal datum plane the following corrections must be applied whether catenary or EDM methods are used for the base measurement.

Slope
If L_1 is the field distance measured at a vertical angle V to the horizontal then a correction C_S is required to reduce L_1 to its horizontal equivalent (L):

$$C_S = -L_1 (1 - \cos V) \tag{4.1a}$$

Alternatively if the heights of the terminals of the line are known $(H_1$ and $H_2)$:

$$C_S = -\left(\frac{(H_1 - H_2)^2}{2L_1} + \frac{(H_1 - H_2)^4}{8L_1^3} \ldots \right) \tag{4.1b}$$

Example 4.1
The slope distance (L_1) measured between A and B is 28.533 m. Determine the correction necessary to reduce the slope distance to the horizontal equivalent (a) if the angle of elevation between A and B is 15°, and (b) if the altitudes of A and B were 785.623 and 793.008 m respectively.

Solution:
(a)
$$C_S = -28.533(1 - \cos 15) = -0.972 \, \text{m}$$
$$L = 28.533 - 0.972 = 27.561 \, \text{m}$$

(b)
$$C_S = -\left(\frac{7.385^2}{2 \times 28.533} \right) + \left(\frac{7.385^4}{8 \times 28.533^3} \right) = -0.972 \, \text{m}$$

Programs 4.1(a), (b), and (c) Slope correction to linear measurements

```
L=28.533    V=15    C=-0.972238398494

100 PRINT "PROGRAM 4.1a"
110 REM THIS PROGRAM DETERMINES THE SLOPE CORRECTION FOR A MEASURED
120 REM DISTANCE AB GIVEN THE ANGLE OF SLOPE
130 INIT
140 SET DEGREES
150 PRINT "ENTER MEASURED DISTANCE; DEGS,MINS,SECS OF SLOPE ANGLE=";
160 INPUT L,D,M,S
170 V=D+M/60+S/3600
180 C=-L*(1-COS(V))
190 PRINT @4:"L=";L;"   ";"V=";V;"   ";"C=";C
200 STOP
210 END
```

```
SLOPE DISTANCE=28.533
ALTITUDE A=785.623   ALTITUDE B=793.008

SLOPE CORRECTION=-0.971709876693
```

```
100 PRINT "PROGRAM 4.1b"
110 REM THIS PROGRAM DETERMINES THE SLOPE CORRECTION FOR A MEASURED
120 REM DISTANCE AB GIVEN THE ALTITUDES OF THE TERMINALS
130 INIT
140 PRINT "ENTER MEASURED DISTANCE, ALTITUDE A, ALTITUDE B=";
150 INPUT L,H1,H2
160 C=-((H1-H2)^2/(2*L)+(H1-H2)^4/(8*L^3))
170 PRINT @4:"SLOPE DISTANCE=";L
180 PRINT @4:"ALTITUDE A=";H1;"   ";"ALTITUDE B=";H2
190 PRINT @4:
200 PRINT @4:"SLOPE CORRECTION=";C
210 STOP
220 END
```

```
SLOPE DISTANCE=28.533   SLOPE ANGLE=15
SLOPE CORRECTION=-0.972238398494
```

```
SLOPE DISTANCE=28.533
ALTITUDE OF A=785.623   ALTITUDE OF B=793.008

SLOPE CORRECTION=0.971709876693
```

```
100 PRINT "PROGRAM 4.1c"
110 REM THIS PROGRAM DETERMINES THE SLOPE CORRECTION FOR A MEASURED
120 REM DISTANCE AB GIVEN EITHER THE ANGLE OF SLOPE OR THE ALTITUDES
130 REM OF THE TERMINALS
140 INIT
150 SET DEGREES
160 PRINT "ENTER MEASURED DISTANCE=";
170 INPUT L
180 PRINT "IF SLOPE ANGLE KNOWN ENTER 0; IF NOT ENTER 1";
190 INPUT Q
200 IF Q=1 THEN 270
210 PRINT "ENTER DEGREES,MINS,SECS OF SLOPE ANGLE=";
220 INPUT D,M,S
230 V=D+M/60+S/3600
240 C=-L*(1-COS(V))
250 PRINT @4:"SLOPE DISTANCE=";L;"   ";"SLOPE ANGLE=";V
260 GO TO 330
270 PRINT "ENTER ALTITUDE OF A, ALTITUDE OF B=";
280 INPUT H1,H2
290 C=(H1-H2)^2/(2*L)+(H1-H2)^4/(8*L^3)
300 PRINT @4:"SLOPE DISTANCE=";L
310 PRINT @4:"ALTITUDE OF A=";H1;"   ";"ALTITUDE OF B=";H2
320 PRINT @4:
330 PRINT @4:"SLOPE CORRECTION=";C
340 STOP
350 END
```

Reduction to sea level

To reduce the measured distance to a common datum (usually mean sea level) a correction C_{MSL} must be applied, the sign of the correction usually being negative:

$$C_{MSL} = -\frac{L_1}{R}\left(\frac{H_1 + H_2}{2}\right) \qquad (4.2)$$

where R = radius of the earth.

Example 4.2
Using the data from Example 4.1 determine the correction to reduce the measured distance to mean sea level given that the radius of the earth is 6378 km.

Solution:

$$C_{MSL} = -\frac{28.533}{6378 \times 10^3} \times \frac{785.623 + 793.008}{2} = -0.004\,m$$

Program 4.2 Sea level correction to linear measurements

```
MEASURED DISTANCE=28.533
ALTITUDE OF A=785.623   ALTITUDE OF B=793.008

MEAN SEA LEVEL CORRECTION=-0.00353112874906


100 PRINT "PROGRAM 4.2"
110 REM THIS PROGRAM DETERMINES THE CORRECTION TO REDUCE A MEASURED
120 REM DISTANCE AB TO DATUM (MEAN SEA LEVEL)
130 PRINT "ENTER MEASURED DISTANCE=";
140 INPUT L
150 PRINT "ENTER ALTITUDE OF A, ALTITUDE OF B=";
160 INPUT H1,H2
170 R=6378000
180 M=(H1+H2)/2
190 C=-(L/R)*M
200 PRINT @4:"MEASURED DISTANCE=";L
210 PRINT @4:"ALTITUDE OF A=";H1;"    ";"ALTITUDE OF B=";H2
212 PRINT @4:
214 PRINT @4:"MEAN SEA LEVEL CORRECTION=";C
220 STOP
230 END
```

4.1.1.1 Catenary corrections

Distances obtained using a tape suspended in catenary may require the following additional corrections: (*a*) standardization, (*b*) temperature, (*c*) tension, (*d*) catenary.

(*a*) *Standardization*
When the length of a tape, determined by comparison with some higher standard, is found to differ from its nominal length, a standardization correction may be applied to the distance given by assuming the tape to be its exact nominal length.

If L_1 = the measured field distance, L_2 = the standardized length of the tape determined under laboratory conditions, L_3 = the nominal length of the tape, then:

$$C_{ST} = (L_3 - L_2)\frac{L_1}{L_2} \tag{4.3}$$

Example 4.3
The calibrated length of a nominal 30 m steel tape is 30.004 m. Determine the standardization correction to be applied directly to measured distances.

Solution:

$$C_{ST} = \left(\frac{30.000 - 30.004}{30.004}\right) \times L_1 = +0.000\,13\,L_1$$

i.e.

$$L = L_1 + 0.000\,13\,L_1$$

Program 4.3(a) Standardization correction

```
NOMINAL LENGTH=30   STANDARD LENGTH=30.004
MEASURED DISTANCE=L1

STANDARDISATION CORRECTION=L1*-1.333155579E-4

100 PRINT "PROGRAM 4.3a"
110 REM THIS PROGRAM DETERMINES THE STANDARDISATION CORRECTION TO BE
120 REM APPLIED TO MEASURED DISTANCES
130 PRINT "ENTER NOMINAL LENGTH OF TAPE, STANDARD LENGTH OF TAPE=";
140 INPUT L3,L2
150 C=(L3-L2)/L2
160 PRINT @4:"NOMINAL LENGTH=";L3;"   ";"STANDARD LENGTH=";L2
170 PRINT @4:"MEASURED DISTANCE=L1"
180 PRINT @4:
190 PRINT @4:"STANDARDISATION CORRECTION=L1*";C
200 STOP
210 END
```

Alternatively the standardization correction may be incorporated with the temperature correction as described in the following section.

(b) Temperature
When the field temperature differs from the temperature prevailing at the time of standardization the temperature correction is:

$$C_T = L_1 K(T_1 - T_2) \tag{4.4}$$

where T_1, T_2 = field temperature and standard temperature respectively, and K = coefficient of linear expansion of tape material.

If, instead of using standard temperature (T_2), a temperature corresponding to the nominal length of the tape is used (T_3), a combined correction for temperature and standardization is obtained:

$$C_{T/ST} = L_1 K (T_1 - T_3) \qquad\qquad (4.5)$$

where

$$T_3 = T_2 - \frac{(L_2 - L_3)}{L_3 K}$$

Example 4.4
The tape referred to in Example 4.3 was calibrated on the flat at 20 °C and 6 kgf tension. Using this tape on the ground the measured distance between two points of the same height was 28.533 m. The standard tension was applied but the prevailing temperature was 16 °C. Determine the corrections to be applied to the measured distances given that the coefficient of expansion of steel is 1.15 × 10⁻⁵ m/°C.

Solution:

$$C_{ST} = +0.000\,13\,L_1 = +0.0037\,\text{m}$$
$$C_T = 28.533 \times (1.15 \times 10^{-5}) \times (16 - 20) = -0.0013\,\text{m}$$
$$\text{Total correction} = C_{ST} + C_T = +0.0024\,\text{m}$$

or, using a combined correction for temperature and standardization:

$$T_3 = 20 - \left(\frac{30.004 - 30.000}{30.000 \times (1.15 \times 10^{-5})} \right) = 8.406$$

$$C_{T/ST} = 28.533 \times (1.15 \times 10^{-5}) \times (16 - 8.4) = +0.0025\,\text{m}$$

Programs 4.3(b) and 4.3(c) Temperature correction, and combined temperature/standardization correction

```
FIELD TEMP=16   STANDARD TEMP=20
MEASURED DISTANCE=28.533

TEMPERATURE CORRECTION=-0.001312518


100 PRINT "PROGRAM 4.3b"
110 REM THIS PROGRAM DETERMINES THE CORRECTION DUE TO TEMPERATURE TO BE
120 REM APPLIED TO A MEASURED DISTANCE AB
130 K=1.15*10^-5
140 PRINT "ENTER MEASURED DISTANCE=";
150 INPUT L
160 PRINT "ENTER STANDARD,FIELD TEMPERATURES=";
170 INPUT T2,T1
180 C=L*K*(T1-T2)
190 PRINT @4:"FIELD TEMP=";T1;"   ";"STANDARD TEMP=";T2
200 PRINT @4:"MEASURED DISTANCE=";L
210 PRINT @4:
220 PRINT @4:"TEMPERATURE CORRECTION=";C
230 STOP
240 END
```

```
NOMINAL LENGTH=30  STANDARD LENGTH=30.004
FIELD TEMP=16  STANDARD TEMP=20
MEASURED DISTANCE=28.533  CORRECTION=0.00249188199991

COMBINED CORRECTION (TEMPERATURE AND STANDARDISATION=0.00249188199991
```

```
100 PRINT "PROGRAM 4.3c"
110 REM THIS PROGRAM DETERMINES THE COMBINED CORRECTION FOR TEMPERATURE
120 REM AND STANDARDISATION TO BE APPLIED TO A MEASURED DISTANCE AB
130 PRINT "ENTER COEFFICIENT OF EXPANSION OF TAPE MATERIAL=";
140 INPUT K
150 PRINT "ENTER MEASURED DISTANCE=";
160 INPUT L
170 PRINT "ENTER STANDARD,NOMINAL LENGTH OF TAPE=";
180 INPUT L2,L3
190 PRINT "ENTER STANDARD,FIELD TEMPERATURES=";
200 INPUT T2,T1
210 T3=T2-(L2-L3)/(L3*K)
220 C=L*K*(T1-T3)
230 PRINT @4:"NOMINAL LENGTH=";L3;"   ";"STANDARD LENGTH=";L2
240 PRINT @4:"FIELD TEMP=";T1;"   ";"STANDARD TEMP=";T2
250 PRINT @4:"MEASURED DISTANCE=";L
260 PRINT @4:
270 PRINT @4:"COMBINED CORRECTION (TEMPERATURE AND STANDARDISATION=";C
280 STOP
290 END
```

(c) Tension
When the tension applied to the tape during field measurement (P_1) differs from that used at standardization (P_2) the correction due to change in tension is

$$C_p = L_1(P_1 - P_2)/AE \qquad (4.6)$$

where A = the cross-sectional area of the tape, and E = Young's modulus of elasticity.

Example 4.5
If the field tension applied in Example 4.4 was 8 kg f determine the required tension correction (width of tape = 3 mm; thickness of tape = 0.25 mm; Young's modulus of elasticity = 20×10^9 kg/m^2).

Solution:

$$C_P = \frac{28.533 \times (8-6)}{(0.003 \times 0.000\,25) \times (20 \times 10^9)} = +0.0038\,\text{m}$$

Program 4.4 Tension correction

```
STANDARD TENSION=6  FIELD TENSION=8
WIDTH TAPE=0.003  THICKNESS TAPE=2.5E-4  AREA=7.5E-7
MEASURED DISTANCE=28.533

TENSION CORRECTION=0.0038044

100 PRINT "PROGRAM 4.4"
```

```
110 REM THIS PROGRAM DETERMINES THE TENSION CORRECTION TO BE APPLIED TO
120 REM A MEASURED DISTANCE AB
130 E=20*10^9
140 PRINT "ENTER STANDARD,FIELD TENSIONS=";
150 INPUT P2,P1
160 PRINT "ENTER WIDTH(M),THICKNESS OF TAPE=";
170 INPUT W,T
180 PRINT "ENTER MEASURED DISTANCE=";
190 INPUT L
200 A=W*T
210 C=L*(P1-P2)/(A*E)
220 PRINT @4:"STANDARD TENSION=";P2;"   ";"FIELD TENSION=";P1
230 PRINT @4:"WIDTH TAPE=";W;"   ";"THICKNESS TAPE=";T;"   ";"AREA=";A
240 PRINT @4:"MEASURED DISTANCE=";L
250 PRINT @4:
260 PRINT @4:"TENSION CORRECTION=";C
270 STOP
280 END
```

(d) Catenary
If the tape has been standardized on the flat, and is then suspended between terminal supports for field measurement, the catenary correction for terminals at the same level is:

$$C_c = - W^2 L_1^3/24P_1^2 \tag{4.7a}$$

where W = weight per unit length of tape, and L_1 = the measured field length under suspension.

If the terminals are not at the same level then:

$$C_c = - W^2 L_1^3 \cos^2 V/24P_1^2 \tag{4.7b}$$

where V = the vertical angle between the terminals.

Example 4.6
If the measurement in Example 4.5 had been made with the tape suspended between the terminals of the line determine the catenary correction given that the tape weighs 1.78 kg.

Solution:

$$C_c = - [(1.78/30)^2 \times 28.533^3]/24 \times 8^2 = - 0.0532 \, \text{m}$$

If the line between the terminals sloped from the horizontal by an angle of, for example, 5° then the catenary correction would be modified as follows:

$$C_c = - 0.0532 \times \cos^2 5° = - 0.0528 \, \text{m}$$

Program 4.5 Catenary correction

```
NOMINAL LENGTH=30   WEIGHT=0.0593333333333   FIELD TENSION= 8
MEASURED DISTANCE=28.533

CATENARY CORRECTION=-0.0532412923388
```

```
100 PRINT "PROGRAM 4.5a"
110 REM THIS PROGRAM DETERMINES THE CATENARY (SAG) CORRECTION TO BE
120 REM APPLIED TO A MEASURED DISTANCE AB, A AND B BEING OF THE SAME
130 REM ALTITUDE
140 PRINT "ENTER MEASURED DISTANCE (SINGLE CATENARY), NOMINAL LENGTH"
150 PRINT "OF TAPE, WEIGHT OF TAPE (KGS),FIELD TENSION=";
160 INPUT L,L3,W,P1
170 W=W/L3
180 C=-(W^2*L^3/(24*P1^2))
190 PRI @4:"NOMINAL LENGTH=";L3;"  ";"WEIGHT=";W;"  ";"FIELD TENSION=";
200 PRINT @4:P1
210 PRINT @4:"MEASURED DISTANCE=";L
220 PRINT @4:
230 PRINT @4:"CATENARY CORRECTION=";C
240 STOP
250 END



NOMINAL LENGTH=30  WEIGHT=0.0593333333333  FIELD TENSION= 8  SLOPE=5
MEASURED DISTANCE=28.533

CATENARY CORRECTION=-0.0528368649072



100 PRINT "PROGRAM 4.5b"
110 REM THIS PROGRAM DETERMINES THE CATENARY CORRECTION TO BE APPLIED TO
120 REM A MEASURED DISTANCE AB OF SLOPE V
130 INIT
140 SET DEGREES
150 PRINT "ENTER MEASURED DISTANCE (SINGLE CATENARY),NOMINAL LENGTH OF"
160 PRINT "TAPE,WEIGHT OF TAPE, FIELD TENSION=";
170 INPUT L,L3,W,P1
180 PRINT "ENTER DEGREES,MINS,SECS OF ANGLE OF SLOPE=";
190 INPUT D,M,S
200 W=W/L3
210 V=D+M/60+S/3600
220 C=-(W^2*L^3*COS(V)^2/(24*P1^2))
230 PRI @4:"NOMINAL LENGTH=";L3;"  ";"WEIGHT=";W;"  ";"FIELD TENSION=";
240 PRINT @4:P1;"  ";"SLOPE=";V
250 PRINT @4:"MEASURED DISTANCE=";L
260 PRINT @4:
270 PRINT @4:"CATENARY CORRECTION=";C
280 STOP
290 END
```

Example 4.7

Four bays of a straight baseline were measured with a steel tape in single catenaries.

Standardization data in respect of the tape was as follows:

Length (on flat) at 20 °C and 6 kg f tension = 30.004 m
Weight = 1.78 kg
Width = 3 mm
Thickness = 0.25 mm

The mean altitude of the baseline was 785 m and the recorded field data is shown in Table 4.1.

Coefficient of expansion of steel = 1.15×10^{-5} m/°C
Young's modulus of elasticity   = 20×10^{9} kg/m^2
Radius of the earth   = 6378 km

Table 4.1   Recorded field data

Bay	Tension (kgf)	Recorded length (m)	Temperature (°C)	Difference in level (m)
1	4	29.643	22	1.06
2	4	30.001	21	1.97
3	4	29.442	21	0.87
4	4	25.728	19	2.56

Determine the length of the base reduced to the horizontal at mean sea level.

Solution:

$$T_N = 20 - \left(\frac{30.004 - 30.000}{30.000 \times (1.15 \times 10^{-5})} \right) = 8.406$$

	C_{TST}	C_P	V	C_C	C_S	Total
Bay 1	+ 0.0046	− 0.0040	2°02′57″	− 0.2385	− 0.0190	− 0.2569
Bay 2	+ 0.0043	− 0.0040	3°45′54″	− 0.2465	− 0.0647	− 0.3109
Bay 3	+ 0.0043	− 0.0040	1°41′36″	− 0.2338	− 0.0129	− 0.2464
Bay 4	+ 0.0031	− 0.0034	5°42′38″	− 0.1546	− 0.1277	− 0.2823

Bay 1 = 29.643 − 0.2569 = 29.3861
  2 = 30.001 − 0.3109 = 29.6901
  3 = 29.442 − 0.2464 = 29.1956
  4 = 25.728 − 0.2823 = 25.4457

Total length   = 113.7175
C_{MSL} = − 0.0140
Total corrected length = 113.7035

Program 4.6 Linear reduction

```
BAY1

CORRECTION DUE TO STANDARDISATION AND TEMP =0.00463418899991
CORRECTION DUE TO TENSION=-0.0039524
CORRECTION DUE TO SLOPE=-0.0189582563472
CORRECTION DUE TO CATENARY=-0.238494164331
TOTAL CORRECTION=-0.256770631678
CORRECTED DISTANCE BAY1=29.3862293683


BAY2

CORRECTION DUE TO STANDARDISATION AND TEMP =0.00434514483324
CORRECTION DUE TO TENSION=-0.00400013333333
CORRECTION DUE TO SLOPE=-0.064749232344
CORRECTION DUE TO CATENARY=-0.246493168875
TOTAL CORRECTION=-0.310897389719
CORRECTED DISTANCE BAY2=29.6901026103
```

BAY3

CORRECTION DUE TO STANDARDISATION AND TEMP =0.00426418299991
CORRECTION DUE TO TENSION=-0.0039256
CORRECTION DUE TO SLOPE=-0.012856891983
CORRECTION DUE TO CATENARY=-0.233770197643
TOTAL CORRECTION=-0.246288506627
CORRECTED DISTANCE BAY3=29.1957114934

BAY4

CORRECTION DUE TO STANDARDISATION AND TEMP =0.00313452799992
CORRECTION DUE TO TENSION=-0.0034304
CORRECTION DUE TO SLOPE=-0.127678431683
CORRECTION DUE TO CATENARY=-0.154598636548
TOTAL CORRECTION=-0.282572940231
CORRECTED DISTANCE BAY4=25.4454270598

CORRECTION DUE TO ALTITUDE=-0.0139962706754
REDUCED DISTANCE FOR LINE=113.703474261


```
100 PRINT "PROGRAM 4.6"
110 INIT
120 REM THIS PROGRAM REDUCES THE CATENARY MEASUREMENTS OF THE BAYS OF A
130 REM SINGLE LINE COMPRISING N BAYS
140 DIM L(5),T(5),P(5),H(5),C(5)
150 A$="CORRECTION DUE TO STANDARDISATION AND TEMP ="
160 B$="CORRECTION DUE TO TENSION="
170 C$="CORRECTION DUE TO SLOPE="
180 D$="CORRECTION DUE TO CATENARY="
190 PRINT "ENTER NOMINAL LENGTH OF TAPE=";
200 INPUT L3
210 PRINT "ENTER STANDARD LENGTH OF TAPE=";
220 INPUT L2
230 PRINT "ENTER STANDARD TEMP=";
240 INPUT T2
250 PRINT "ENTER STANDARD TENSION="
260 INPUT P2
270 PRINT "ENTER AVERAGE ALTITUDE=";
280 INPUT Z
290 PRINT "ENTER WIDTH OF TAPE (M),THICKNESS OF TAPE (M)=";
300 INPUT B1,B2
310 A=B1*B2
320 PRINT "ENTER WEIGHT OF TAPE (KG)=";
330 INPUT W
340 W=W/L3
350 T3=T2-(L2-L3)/(L3*11.5*10^-6)
360 SET DEGREES
370 S=0
380 PRINT "ENTER NUMBER OF BAYS OF LINE=";
390 INPUT N
400 FOR I=1 TO N
410 PRINT "ENTER MEASURED LENGTH OF BAY";I;"=";
420 INPUT L(I)
430 PRINT "ENTER FIELD TEMP OF BAY";I;"=";
440 INPUT T(I)
450 PRINT "ENTER FIELD TENSION OF BAY";I;"=";
460 INPUT P(I)
470 PRINT "ENTER DIFFERENCE IN ALTITUDE OF TERMINALS OF BAY";I;"=";
480 INPUT H(I)
490 Q1=0
500 Q1=Q1+L(I)*((T(I)-T3)*(11.5*10^-6))
510 Q2=0
520 Q2=Q2+L(I)*(P(I)-P2)/(A*(20*10^9))
530 Q3=0
540 Q3=Q3-(H(I)^2/(2*L(I))+H(I)^4/(8*L(I)^3))
550 V=ATN(H(I)/L(I))
560 Q4=0
570 Q4=Q4-W^2*L(I)^3*COS(V)^2/(24*P(I)^2)
```

```
580 C(I)=0
590 C(I)=C(I)+Q1+Q2+Q3+Q4
600 S=S+(L(I)+C(I))
610 PRINT @4:"BAY";I
620 PRINT @4:
630 PRINT @4:A$;Q1
640 PRINT @4:B$;Q2
650 PRINT @4:C$;Q3
660 PRINT @4:D$;Q4
670 PRINT @4:"TOTAL CORRECTION=";C(I)
680 PRINT @4:"CORRECTED DISTANCE BAY";I;"=";L(I)+C(I)
690 PRINT @4:
700 PRINT @4:
710 NEXT I
720 Q5=0
730 Q5=Q5-S*Z/(6378*10^3)
740 PRINT @4:"CORRECTION DUE TO ALTITUDE=";Q5
750 S=S+Q5
760 PRINT @4:"REDUCED DISTANCE FOR LINE=";S
770 STOP
780 END
```

4.1.1.2 Corrections to distances derived from EDM

The value of the distance as presented in the digital readout of most EDM instruments requires the application of instrumental corrections and corrections for the variation in refractive index in addition to those referred to in Section 4.1.1.

(a) *Instrumental corrections*
These are constants for individual instruments and include zero and cyclic error corrections determined by calibration.

(b) *Variation in refractive index*
The velocity of propagation of the electromagnetic signal is related to the refractive index which depends on the frequency of the transmitted signal and the prevailing atmospheric conditions.

If D_1 is the distance as recorded by the instrument corresponding with a standard refractive index N_2 then the corrected distance D_2 is given by:

$$D_2 = \frac{D_1 N_2}{N_1}$$

where N_1 is the refractive index based on prevailing atmospheric conditions.

The correction required to reduce the recorded distance is therefore:

$$C_R = \left(\frac{N_2}{N_1} - 1\right)D_1 \qquad (4.8)$$

For light waves:

$$N_1 = 1 + \frac{N_3 - 1}{1 + CT} \frac{P}{760} \tag{4.9}$$

where N_3 = the group refractive index, C = coefficient of expansion of air (3.663×10^{-3} approximately), T = prevailing atmospheric temperature (°C), P = prevailing atmospheric pressure (mm Hg).

For microwaves:

$$(N_1 - 1)10^6 = \frac{103.49}{273 + T}(P - E) + \frac{86.26}{273 + T}\left(1 + \frac{5748}{273 + T}\right)E \tag{4.10}$$

where E = partial atmospheric water vapour pressure, and:

$$E = E' - 0.000\,66\,P(T - T') \tag{4.11}$$

where E' = saturation vapour pressure of the atmosphere at T' (mm Hg), T = mean dry bulb temperature (°C), and T' = mean wet bulb temperature (°C).

E' may be calculated from:

$$\log_{10} E' = 0.660\,887 + 3.154\,882\,(T'/100)$$
$$- 1.274\,528\,(T'/100)^2 + 0.375\,114\,(T'/100)^3 \tag{4.12}$$

Example 4.8
Determine the refraction correction to be applied to distances recorded by a light-wave instrument given the following data:

Group refractive index   = 1.000 294 1 @ 0 °C and 760 mm Hg
Standard refractive index = 1.000 281 7 @ 0 °C and 760 mm Hg
Prevailing temperature   = 20 °C
Prevailing pressure   = 730 mm Hg

Solution:

$$N_1 = 1 + \frac{1.000\,294\,1 - 1}{1 + (3.663 \times 10^{-3} \times 20)} \times \frac{730}{760} = 1.000\,263\,2$$

$$C_R = \left(\frac{1.000\,281\,7}{1.000\,263\,2} - 1\right)D_1$$

$$= (1.85 \times 10^{-5})D_1$$
$$= +1.85\,\text{mm per }100\,\text{m}$$

Program 4.7(a) EDM refraction correction (1)

```
GROUP REFRACTIVE INDEX=1.0002941
STANDARD REFRACTIVE INDEX=1.0002817
PREVAILING TEMP=20PREVAILING PRESSURE=730

CORRECTION =1.84869743833MM PER 100 M


100 PRINT "PROGRAM 4.7a"
110 REM THIS PROGRAM DETERMINES THE REFRACTION CORRECTION FOR USE WITH
120 REM LIGHT WAVE EDM INSTRUMENTS
130 INIT
140 C=3.663*10^-3
150 PRINT "ENTER GROUP REFRACTIVE INDEX";
160 INPUT N3
170 PRINT "ENTER STANDARD REFRACTIVE INDEX,STANDARD TEMP(C),STANDARD"
180 PRINT "PRESSURE(MM)";
190 INPUT N2,T,P
200 PRINT "ENTER PREVAILING TEMP(C),PREVAILING PRESSURE(MM)";
210 INPUT T1,P1
220 N1=1+(N3-1)/(1+C*T1)*(P1/760)
230 C1=N2/N1-1
240 PRINT @4:"GROUP REFRACTIVE INDEX=";N3
250 PRINT @4:"STANDARD REFRACTIVE INDEX=";N2
260 PRINT @4:"PREVAILING TEMP=";T1;"PREVAILING PRESSURE=";P1
270 PRINT @4:
280 PRINT @4:"CORRECTION =";C1*10^5;"MM PER 100 M"
290 STOP
300 END
```

Example 4.9
The distance recorded on a microwave instrument is 11 260.522 m. The instrument's standard refractive index is 1.000 325 and the prevailing atmospheric conditions are as follows:

Temperature (dry bulb) = 3.3 °C
(wet bulb) = 2.3 °C
Pressure   = 977 mb

Determine the distance corrected for refraction.

Solution:

$$\log_{10} E' = 0.660\,887 + 3.154\,882\,(2.3/100)$$
$$- 1.274\,528\,(2.3/100)^2 + 0.375\,114\,(2.3/100)^3$$
$$= 0.732\,779\,6$$
$$E' = 5.405$$
$$E = 5.405 - 0.000\,66\,(977 \times 0.750\,062)(1)$$
$$= 4.921$$

$$(N_1 - 1)10^6 = \frac{103.49}{273 + 3.3}[(977 \times 0.750\,062) - 4.921]$$
$$+ \frac{86.26}{273 + 3.3}\left(1 + \frac{5748}{273 + 3.3}\right)4.921$$
$$= 272.636 + 33.497$$
$$= 306.133$$

$$N_1 = \frac{306.113 + 10^6}{10^6} = 1.000\,306$$

$$D_2 = 11\,260.522 \times \frac{1.000\,325}{1.000\,306} = 11\,260.735\,\text{m}$$

Program 4.7(b) EDM refraction correction (2)

```
STANDARD REFRACTIVE INDEX=1.000325
PREVAILING DRY TEMP=3.3PREVAILING WET TEMP=2.3PREVAILING PRESSURE=732.810574
PARTIAL WATER VAPOUR PRESSURE=4.92114498145
MEASURED DISTANCE=11260.522

CORRECTED DISTANCE=11260.7343767


100 PRINT "PROGRAM 4.7b"
110 REM THIS PROGRAM DETERMINES THE REFRACTION CORRECTION FOR USE WITH
120 REM MICRO WAVE EDM INSTRUMENTS
130 PRINT "ENTER STANDARD REFRACTIVE INDEX";
140 INPUT N2
150 PRINT "ENTER PREVAILING DRY TEMP,PREVAILING WET TEMP,PREVAILING"
160 PRINT "PRESSURE";
170 INPUT T1,T2,P1
180 P1=P1*0.750062
190 PRINT "ENTER MEASURED DISTANCE";
200 INPUT D1
210 E=0.660887+3.154882*T2/100-1.274528*(T2/100)^2
220 E=E+0.375114*(T2/100)^3
230 E1=10^E
240 E2=E1-6.6E-4*P1*(T1-T2)
250 R=103.49/(273+T1)*(P1-E2)
260 S=86.26/(273+T1)*((1+5748/(273+T1))*E2)
270 N1=(R+S)/10^6+1
280 D2=D1*N2/N1
290 PRINT @4:"STANDARD REFRACTIVE INDEX=";N2
300 PRINT @4:"PREVAILING DRY TEMP=";T1;"PREVAILING WET TEMP=";T2;
310 PRINT @4:"PREVAILING PRESSURE=";P1
320 PRINT @4:"PARTIAL WATER VAPOUR PRESSURE=";E2
330 PRINT @4:"MEASURED DISTANCE=";D1
335 PRINT @4:
340 PRINT @4:"CORRECTED DISTANCE=";D2
360 STOP
370 END
```

4.1.2 Triangulation adjustment

Prior to computation of the coordinates of points in a triangulation network, random errors in the angular observations must be distributed so as to make the figures comprising the network geometrically consistent. This adjustment is best made using the method of least squares but if high accuracy is not sought, simple and more approximate adjustment procedures may be used.

When the triangulation comprises a series of adjoining triangles of limited extent the only condition to be satisfied in the adjustment is that the angles of each individual triangle must sum to 180°. This is achieved simply by distributing the angular misclosure equally among the three angles of the triangle.

When the network comprises geometrical figures other than individual triangles further conditions must be satisfied. The method of *equal shifts* satisfies such additional conditions by distributing the angular errors uniformly between the constituent angles in such a way that the computed length of any side will have the same value irrespective of which route is used in the computation.

4.1.2.1 *Number of conditions*

The number of conditions to be satisfied in a network adjustment for a network with a single fixed side may be determined from the following formulae:

Number of angle conditions $= L - L_1 - (S - S_1) + 1 + C$ (4.9*c*)
Number of side conditions $= L - 2S + 3$ (4.9*b*)
Total number of conditions $= N - 2S + 4$ (4.9*c*)

where $S =$ total number of stations, $S_1 =$ number of unoccupied stations, $L =$ total number of lines, including those observed one way only, $L_1 =$ number of lines observed one way only, $C =$ number of points at which the observed angles must sum to $360°$, and $N =$ total number of observed angles.

Additional conditions are required if other datum values are to be held fixed. For instance, each additional fixed side will require an additional side condition and each fixed angle between two adjacent sides of fixed length and position will require an additional angle condition.

Example 4.8
Figure 4.1 illustrates the application of the above formulae to five variations of the same basic figure used to fix stations C and D from a base AB.

For Figure 4.1(*a*):

Number of angle conditions $= 5 - 3 - (4 - 1) + 1 + 0 = 0$
Number of side conditions $= 5 - 8 + 3$ $= 0$
Total number of conditions $= 4 - 8 + 4$ $= 0$

For Figure 4.1(*b*):

Number of angle conditions $= 6 - 2 - (4 - 0) + 1 + 0 = 1$
Number of side conditions $= 6 - 8 + 3$ $= 1$
Total number of conditions $= 6 - 8 + 4$ $= 2$

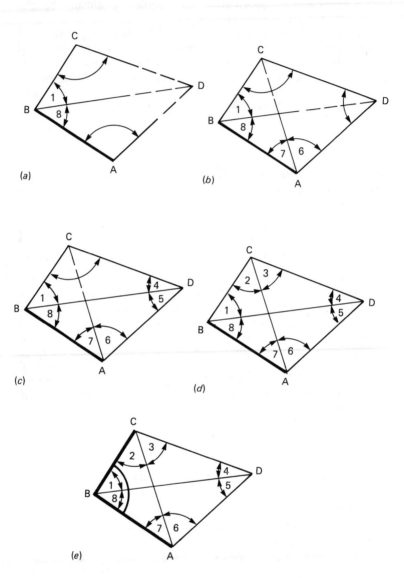

Figure 4.1 Triangulation adjustment: variations in angle observations in respect of a quadrilateral extension from base **AB**

For Figure 4.1(c):

Number of angle conditions $= 6 - 1 - (4 - 0) + 1 + 0 = 2$
Number of side conditions $\quad = 6 - 8 + 3 \qquad\qquad\quad = 1$
Total number of conditions $\; = 7 - 8 + 4 \qquad\qquad\quad = 3$

For Figure 4.1(d):

Number of angle conditions $= 6 - 0 - (4 - 0) + 1 + 0 = 3$
Number of side conditions $\quad = 6 - 8 + 3 \qquad\qquad\quad = 1$
Total number of conditions $\; = 8 - 8 + 4 \qquad\qquad\quad = 4$

For Figure 4.1(e):
If the side BC also formed part of the existing network and was fixed in length and position relative to AB, an additional side condition would be required together with an additional angle condition (the sum of the angles 1 and 8 must be made to equal the fixed value for the angle ABC).

Number of angle conditions $= 6 - 0 - (4 - 0) + 1 + 0$
$\qquad\qquad\qquad\qquad\qquad\quad$ (+ one fixed angle ABC) $\quad = 4$
Number of side conditions $\quad = 6 - 8 + 3$ (+ one fixed side BC) $= 2$
Total number of conditions $= 8 - 8 + 4$
$\qquad\qquad\qquad\qquad$ (+ one fixed angle and one fixed side) $\quad = 6$

Example 4.9
Figure 4.2 shows a triangulation network based on the side FG comprising a single triangle ABG, and a braced quadrilateral BHFG interlocking with a centre point polygon H(BCDEF).

For the whole figure:

Number of angle conditions $= 15 - 0 - (8 - 0) + 1 + 1 = 9$
Number of side conditions $\quad = 15 - 16 + 3 \qquad\qquad = 2$
Total number of conditions $= 23 - 16 + 4 \qquad\qquad = 11$

Programs 4.8–4.10 Number of conditions in triangulation network

```
EXAMPLE 4.8                        SCHEME4
                                   NUMBER OF ANGLE CONDITIONS=3
SCHEME1                            NUMBER OF SIDE CONDITIONS=1
NUMBER OF ANGLE CONDITIONS=0       TOTAL NUMBER OF CONDITIONS=4
NUMBER OF SIDE CONDITIONS=0
TOTAL NUMBER OF CONDITIONS=0
                                   SCHEME5
                                   NUMBER OF ANGLE CONDITIONS=4
SCHEME2                            NUMBER OF SIDE CONDITIONS=2
NUMBER OF ANGLE CONDITIONS=1       TOTAL NUMBER OF CONDITIONS=6
NUMBER OF SIDE CONDITIONS=1
TOTAL NUMBER OF CONDITIONS=2


SCHEME3                            EXAMPLE 4.9
NUMBER OF ANGLE CONDITIONS=2       NUMBER OF ANGLE CONDITIONS=9
NUMBER OF SIDE CONDITIONS=1        NUMBER OF SIDE CONDITIONS=2
TOTAL NUMBER OF CONDITIONS=3       TOTAL NUMBER OF CONDITIONS=11
```

```
100 PRINT "PROGRAM 4.8 - 4.10"
110 REM THIS PROGRAM DETERMINES THE NUMBER OF CONDITIONS REQUIRED FOR
120 REM THE ADJUSTMENT OF THE TRIANGULATION SCHEMES REFERRED TO IN
130 REM EXAMPLES 4.8 AND 4.9
140 INIT
150 A$="NUMBER OF ANGLE CONDITIONS="
160 B$="NUMBER OF SIDE CONDITIONS="
170 C$="TOTAL NUMBER OF CONDITIONS="
180 PRINT @4:"EXAMPLE 4.8"
190 PRINT @4:
200 PRINT "ENTER NUMBER OF SCHEMES=";
210 INPUT N
220 FOR I=1 TO N
230 PRINT "ENTER TOTAL NUMBER OF STATIONS IN NETWORK";I;"=";
240 INPUT S
250 PRINT "ENTER NUMBER OF UNOCCUPIED STATIONS IN NETWORK";I;"=";
260 INPUT S1
270 PRINT "ENTER TOTAL NUMBER OF LINES IN NETWORK";I;"=";
280 INPUT L
290 PRI "ENTER NUMBER OF LINES OBSERVED ONE WAY ONLY IN NETWORK";I;"=";
300 INPUT L1
310 PRINT "ENTER NUMBER OF 'CENTRE POINTS' IN NETWORK";I;"=";
320 INPUT C
330 PRINT "ENTER TOTAL NUMBER OF OBSERVED ANGLES IN NETWORK";I;"=";
340 INPUT O
350 PRINT "ENTER NUMBER OF ADDITIONAL FIXED SIDES IN NETWORK";I;"=";
360 INPUT F1
370 PRINT "ENTER NUMBER OF ADDITIONAL FIXED ANGLES IN NETWORK";I;"=";
380 INPUT F2
390 A1=0
400 A1=A1+L-L1-(S-S1)+C+F2+1
410 A2=0
420 A2=A2+L-2*S+F1+3
430 T=0
440 T=T+O-2*S+F1+F2+4
450 IF I=N THEN 530
460 PRINT @4:"SCHEME";I
470 PRINT @4:A$;A1
480 PRINT @4:B$;A2
490 PRINT @4:C$;T
500 PRINT @4:
510 PRINT @4:
520 NEXT I
530 PRINT @4:
540 PRINT @4:
550 PRINT @4:"EXAMPLE 4.9"
560 PRINT @4:A$;A1
570 PRINT @4:B$;A2
580 PRINT @4:C$;T
590 STOP
600 END
```

4.1.2.2 *Equal shift adjustment*

Referring to Figure 4.2 and Example 4.9, assuming a free network with the single base FG, the 11 conditions would be satisfied as follows:

(*a*) Triangle ABG (3 sides, 3 angles, 3 stations):

Number of angle conditions $= 3 - 3 + 1 = 1$
Number of side conditions $\ = 3 - 6 + 3 = 0$
Total number of conditions $= 3 - 6 + 4 = 1$
Angle condition: $1 + 2 + 17 = 180°$.

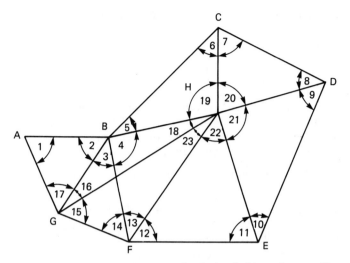

Figure 4.2 Triangulation adjustment: triangle, interlocking polygon and braced quadrilateral

(b) Braced quadrilateral BHFG (6 sides, 8 angles, 4 stations):

Number of angle conditions $= 6 - 4 + 1 = 3$
Number of side conditions $= 6 - 8 + 3 = 1$
Total number of conditions $= 8 - 8 + 4 = 4$
Angle conditions:

$$3 + 4 + 18 + 23 + 13 + 14 + 15 + 16 = 360°$$
$$4 + 18 - (14 + 15) = 0$$
$$3 + 16 - (13 + 23) = 0$$

Side condition:

$$\frac{\sin 4 \times \sin 23 \times \sin 14 \times \sin 16}{\sin 3 \times \sin 18 \times \sin 13 \times \sin 15} = 1 \qquad (4.10a)$$

or

$$\log \sin 4 + \log \sin 23 + \log \sin 14 + \log \sin 16$$
$$- \log \sin 3 - \log \sin 18 - \log \sin 13 - \log \sin 15 = 0 \qquad (4.10b)$$

The angles used in the side condition are the observed angles corrected for the angle conditions.

The corrections arising out of the side condition are all numerically equal (hence the name 'equal shift') and must be applied with alternate positive and negative signs to successive angles in order that the finally adjusted angles continue to satisfy the angle conditions.

The side condition correction is

$$C'' = \frac{D}{d} \qquad\qquad (4.11)$$

where $D =$ the difference between the two sides of the logarithmic side equation (Equation (4.10b)), $d =$ the difference in the logarithmic sin for $1''$ of arc in respect of each angle ($= 10^7 \cot A / 206\,265$ where A is the angle adjusted for the angle condition).

(c) Centre point polygon H(BCDEF) (10 sides, 16 angles, 6 stations):

Number of angle conditions $= 10 - 6 + 1 + 1 = 6$ (including a centre point angle condition)
Number of side conditions $= 10 - 2 + 3 = 1$
Total number of conditions $= 15 - 12 + 4 = 7$

However, the condition that the angles of the triangle BFH sum $180°$ has already been satisfied in the adjustment of the quadrilateral. There are therefore 5 angle conditions and 1 side condition to be satisfied:

Angle conditions:
$5 + 6 + 19 = 180°$
$7 + 8 + 20 = 180°$
$9 + 10 + 21 = 180°$
$11 + 12 + 22 = 180°$
$19 + 20 + 21 + 22 = 360° - (18 + 23)$

Side condition:
$$\frac{\sin 5 \times \sin 7 \times \sin 9 \times \sin 11 \times \sin 13}{\sin 6 \times \sin 8 \times \sin 10 \times \sin 12 \times \sin 4} = 1$$

or

$\log \sin 5 + \log \sin 7 + \log \sin 9 + \log \sin 11$
$+ \log \sin 13 - \log \sin 6 - \log \sin 8 - \log \sin 10$
$- \log \sin 12 - \log \sin 4 = 0$

In this example no adjustment arising out of the side condition is made to angles 4 and 13.

Example 4.10
Figure 4.3 shows a braced quadrilateral with its observed angles. Table 4.2 shows the adjusted angles, determined in Program 4.11

Table 4.2 Solution to Example 4.10

1 Angle No	2 Observed °	′	″	3 Sum of opposites °	′	″	4 First adjustment to 360° ″	5 to opps ″	6 First adjusted angles °	′	″	7 Log sines odd nos	8 Log sines even nos	9 d	10 Second adjustment angle (C″)	11 Final adjusted angles °	′	″
1	20	41	10	60	08	15	−0.5	+4.2	20	41	13.7	−0.451 8996		+56	−1.5	20	41	12.2
2	39	27	05				−0.5	+4.3	39	27	08.8		−0.196 9271	+26	+1.5	39	27	10.3
3	71	58	38	119	51	44	−0.5	−2.8	71	58	34.7	−0.021 8521		+7	−1.5	71	58	33.2
4	47	53	06				−0.5	−2.7	47	53	02.8		−0.129 7191	+19	+1.5	47	53	04.3
5	29	02	53	60	08	32	−0.5	−4.2	29	02	48.3	−0.313 7901		+37	−1.5	29	02	46.8
6	31	05	39				−0.5	−4.3	31	05	34.2		−0.286 9917	+35	+1.5	31	05	35.7
7	78	46	31	119	51	33	−0.5	+2.8	78	46	33.3	−0.008 3970		+5	−1.5	78	46	31.8
8	41	05	02				−0.5	+2.7	41	05	04.2		−0.182 3214	+25	+1.5	41	05	05.7
Sum	360	00	04	360	00	04	−4.0	0.0	360	00	00.0	−0.795 9288	−0.795 9593	+210	0.0	360	00	00.0

$$D = \frac{9288}{210\,)\,305} = 1.452$$

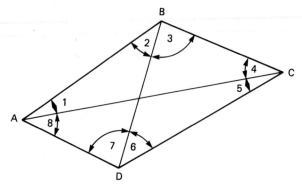

Figure 4.3 Triangulation adjustment: braced quadrilateral

	°	′	″		°	′	″
1.	20	41	10	5.	29	02	53
2.	39	27	05	6.	31	05	39
3.	71	58	38	7.	78	46	31
4.	47	53	06	8.	41	05	02

Program 4.11 Quadrilateral adjustment

```
OBSERVED ANGLES
ANGLE1=20 41 10
ANGLE2=39 27 5
ANGLE3=71 58 38
ANGLE4=47 53 6
ANGLE5=29 2 53
ANGLE6=31 5 39
ANGLE7=78 46 31
ANGLE8=41 5 2
SUM OF ANGLES=360 0 4
FIRST ANGLE MISCLOSURE=0 0 4.49999997899
SUM OF ANGLES FOLLOWING FIRST ADJUSTMENT=360 0 0
SECOND ANGLE MISCLOSURE=0 0 17.5000000007
THIRD ANGLE MISCLOSURE=0 0 11.4999999995
SUM OF ANGLES FOLLOWING 2ND AND 3RD ANGLE ADJUSTMENT=360 0 0
SUM OF LOG SIN ODD ANGLES=-0.795928669699
SUM OF LOG SIN EVEN ANGLES=-0.795959223248
DIFFERENCE IN SUMS OF LOG SINS=3.055354892E-5
SUM OF DIFF FOR ONE SEC=208.380898028
SIDE CORRECTION=4.072876642E-4
ADJUSTED ANGLES

ANGLE1=20 41 12

ANGLE2=39 27 10

ANGLE3=71 58 33

ANGLE4=47 53 4

ANGLE5=29 2 47
```

```
ANGLE6=31 5 36

ANGLE7=78 46 32

ANGLE8=41 5 6
SUM OF FINAL ADJUSTED ANGLES =360 0 0


100 PRINT "PROGRAM 4.11"
110 INIT
120 REM THIS PROGRAM PERFORMS AN EQUAL SHIFT ADJUSTMENT OF A BRACED
130 REM QUADRILATERAL
140 SET DEGREES
150 DIM A(8),T(8),D(8),M(8),S(8),Z(5),J(5),L(5)
160 PRINT
170 PRINT @4:"OBSERVED ANGLES"
180 Z(1)=0
190 FOR I=1 TO 8
200 PRINT "ENTER DEGREES,MINS,SECS OF ANGLE A";I;"=";
210 INPUT D(I),M(I),S(I)
220 PRINT @4:"ANGLE";I;"=";D(I);M(I);S(I)
230 A(I)=D(I)+M(I)/60+S(I)/3600
240 Z(1)=Z(1)+A(I)
250 NEXT I
260 O=Z(1)
270 GOSUB 1410
280 PRINT
290 PRINT "SUM OF ANGLES=";O
300 PRINT @4:"SUM OF ANGLES=";D(1);INT(M(1));INT(S(1))
310 B=Z(1)-360
320 O=ABS(B)
330 GOSUB 1410
340 D(1)=D(1)*SGN(B)
350 PRINT "FIRST ANGLE MISCLOSURE=";O
360 PRINT @4:"FIRST ANGLE MISCLOSURE=";D(1);INT(M(1));S(1)
370 C=-(B/8)
380 PRINT "FIRST ANGLE CORRECTION=";C
390 O=ABS(C)
400 GOSUB 1410
410 D(1)=D(1)*SGN(C)
420 PRINT
430 FOR I=1 TO 8
440 A(I)=A(I)+C
450 PRINT "A";I;"=";A(I)
460 NEXT I
470 PRINT
480 Z(2)=A(1)+A(2)+A(3)+A(4)+A(5)+A(6)+A(7)+A(8)
490 O=Z(2)
500 GOSUB 1410
510 PRINT "SUM OF ANGLES=";O
520 PRINT @4:"SUM OF ANGLES FOLLOWING FIRST ADJUSTMENT=";D(1);INT(M(1));
530 PRINT @4:INT(S(1))
540 E=A(1)+A(2)
550 F=A(6)+A(5)
560 G=A(3)+A(4)
570 H=A(7)+A(8)
580 J(1)=E-F
590 O=ABS(J(1))
600 GOSUB 1410
610 PRINT "SECOND ANGLE MISCLOSURE=";O
620 PRINT @4:"SECOND ANGLE MISCLOSURE=";D(1);INT(M(1));S(1)
630 K=-(J(1)/4)
640 PRINT "SECOND ANGLE CORRECTION=";K
650 PRINT
660 A(1)=A(1)+K
670 A(2)=A(2)+K
680 A(5)=A(5)-K
690 A(6)=A(6)-K
700 PRINT
710 J(2)=A(1)+A(2)-(A(5)+A(6))
720 PRINT "J(2)=";J(2)
730 L(1)=G-H
```

```
740 O=ABS(L(1))
750 GOSUB 1410
760 D(1)=D(1)*SGN(L(1))
770 PRINT
780 PRINT "THIRD ANGLE MISCLOSURE =";O
790 PRINT @4;"THIRD ANGLE MISCLOSURE=";D(1);INT(M(1));S(1)
800 N=-(L(1)/4)
810 PRINT "THIRD ANGLE CORRECTION =";N
820 PRINT
830 A(3)=A(3)+N
840 A(4)=A(4)+N
850 A(7)=A(7)-N
860 A(8)=A(8)-N
870 PRINT
880 L(2)=A(3)+A(4)-(A(7)+A(8))
890 PRINT "L(2)=";L(2)
900 Z(2)=A(1)+A(2)+A(3)+A(4)+A(5)+A(6)+A(7)+A(8)
910 O=Z(2)
920 GOSUB 1410
930 PRI @4;"SUM OF ANGLES FOLLOWING 2ND AND 3RD ANGLE ADJUSTMENT=";D(1);
940 PRINT @4;INT(M(1));INT(S(1))
950 PRINT
960 P=LGT(SIN(A(1)))+LGT(SIN(A(3)))+LGT(SIN(A(5)))+LGT(SIN(A(7)))
970 PRINT "SUM OF LOG SIN ODDS =";P
980 PRINT @4;"SUM OF LOG SIN ODD ANGLES=";P
990 Q=LGT(SIN(A(2)))+LGT(SIN(A(4)))+LGT(SIN(A(6)))+LGT(SIN(A(8)))
1000 PRINT "SUM OF LOG SIN EVENS =";Q
1010 PRINT @4;"SUM OF LOG SIN EVEN ANGLES=";Q
1020 R=P-Q
1030 PRINT @4;"DIFFERENCE IN SUMS OF LOG SINS=";R
1040 R=ABS((P-Q)*10^7)
1050 PRINT
1060 U=0
1070 FOR I=1 TO 8
1080 T(I)=21.055*(1/TAN(A(I)))
1090 PRINT "DIFF FOR ONE SEC,A";I;"=";T(I)
1100 U=U+T(I)
1110 NEXT I
1120 PRINT
1130 PRINT "SUM OF DIFF FOR ONE SEC =";U
1140 PRINT @4;"SUM OF DIFF FOR ONE SEC=";U
1150 PRINT
1160 V=ABS(R/U/3600)
1170 PRINT "SIDE CORRECTION =";V
1180 PRINT @4;"SIDE CORRECTION=";V
1190 PRINT
1200 A(1)=A(1)-V
1210 A(2)=A(2)+V
1220 A(3)=A(3)-V
1230 A(4)=A(4)+V
1240 A(5)=A(5)-V
1250 A(6)=A(6)+V
1260 A(7)=A(7)-V
1270 A(8)=A(8)+V
1280 PRINT @4;"ADJUSTED ANGLES"
1290 FOR I=1 TO 8
1300 D(I)=INT(A(I))
1310 M(I)=(A(I)-D(I))*60
1320 S(I)=(M(I)-INT(M(I)))*60+0.5
1330 PRINT @4;
1340 PRINT @4;"ANGLE";I;"=";D(I);INT(M(I));INT(S(I))
1350 NEXT I
1360 Z(2)=A(1)+A(2)+A(3)+A(4)+A(5)+A(6)+A(7)+A(8)
1370 Z(2)=0
1380 GOSUB 1410
1390 PRINT @4;"SUM OF FINAL ADJUSTED ANGLES =";D(1);INT(M(1));INT(S(1))
1400 GO TO 1450
1410 D(1)=INT(O)
1420 M(1)=(O-D(1))*60
1430 S(1)=(M(1)-INT(M(1)))*60+0.5
1440 RETURN
1450 STOP
1460 END
```

Example 4.11
Figure 4.4 shows a centre-point polygon the observed angles of which are as follows:

	°	′	″		°	′	″
1.	39	22	53	7.	117	10	38
2.	23	26	32	8.	132	05	56
3.	12	21	21	9.	110	43	28
4.	35	32	37				
5.	44	49	35				
6.	24	26	50				

Determine the adjusted angles. The solution is shown in Table 4.3.

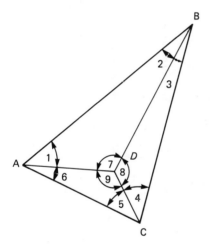

Figure 4.4 Triangulation adjustment: centre-point polygon

Table 4.3 Solution to Example 4.11

1 Angle No	2 Observed ° ' "	3 Correction (180°) "	4 Central angles "	5 Correction (360°) "	6 First adjusted angles ° ' "	7 Log sines odd numbers	8 Log sines even numbers	9 d	10 Second adjustment angles (C")	11 Final adjusted angles ° ' "
1	39 22 53	−1		+1	39 22 53	−0.1975824		25.6	−0.6	39 22 52.4
2	23 26 32	−1		+1	23 26 32		−0.4003089	48.6	+0.6	23 26 32.6
7	117 10 38	−1	37	−2	117 10 35					117 10 35.0
Sum	180 00 03	−3			180 00 00					180 00 00
3	12 21 21	+2		+1	12 21 24	−0.6695931		96.1	−0.6	12 21 23.4
4	35 32 37	+2		+1	35 32 40		−0.2355741	29.5	+0.6	35 32 40.6
8	132 05 56	+2	58	−2	132 05 56					132 05 56.0
Sum	179 59 54	+6			180 00 00					180 00 00
5	44 49 35	+2		+1	44 49 38	−0.1518286		21.2	−0.6	44 49 37.4
6	24 26 50	+2		+1	24 26 53		−0.3831380	46.3	+0.6	24 26 53.6
9	110 43 28	+3	31	−2	110 43 29					110 43 29.0
					180 00 00					180 00 00
Sum	179 59 53	+7	+06		180 00 00	−1.0190041	−1.0190210	267.3		180 00 00

$$-1.0190041$$
$$\frac{169}{267.3} = 0.6''$$

Program 4.12 Centre-point polygon adjustment

```
OBSERVED ANGLE1=39 22 53
OBSERVED ANGLE2=23 26 32
OBSERVED ANGLE3=12 21 21
OBSERVED ANGLE4=35 32 37
OBSERVED ANGLE5=44 49 35
OBSERVED ANGLE6=24 26 50
OBSERVED ANGLE7=117 10 38
OBSERVED ANGLE8=132 5 56
OBSERVED ANGLE9=110 43 28
SUM OF ANGLES OF TRIANGLE1=180.000833333
MISCLOSURE TRIANGLE1=0 0 3.49999999406
FIRST ADJUSTED VALUE ANGLE1=39 22 52
FIRST ADJUSTED VALUE ANGLE2=23 26 31
FIRST ADJUSTED VALUE ANGLE7= 117 10 37
SUM OF ANGLES OF TRIANGLE2=179.998333333
MISCLOSURE TRIANGLE2=0 0 6.5000000045
FIRST ADJUSTED VALUE ANGLE3=12 21 23
FIRST ADJUSTED VALUE ANGLE4=35 32 39
FIRST ADJUSTED VALUE ANGLE8= 132 5 58
SUM OF ANGLES OF TRIANGLE3=179.998055556
MISCLOSURE TRIANGLE3=0 0 7.50000000579
FIRST ADJUSTED VALUE ANGLE5=44 49 37
FIRST ADJUSTED VALUE ANGLE6=24 26 52
FIRST ADJUSTED VALUE ANGLE9= 110 43 30

SUM OF CENTRE POINT ANGLES=360.001481481
MISCLOSURE CENTRE POINT ANGLES =0 0 5.83333332278
SECOND ADJUSTED VALUES ANGLE7=117 10 35
SECOND ADJUSTED VALUES ANGLE8=132 5 56
SECOND ADJUSTED VALUES ANGLE9=110 43 29
SECOND ADJUSTED VALUES ANGLE1=39 22 53
SECOND ADJUSTED VALUES ANGLE2=23 26 32
SECOND ADJUSTED VALUES ANGLE3=12 21 24
SECOND ADJUSTED VALUES ANGLE4=35 32 40
SECOND ADJUSTED VALUES ANGLE5=44 49 38
SECOND ADJUSTED VALUES ANGLE6=24 26 53
SUM OF ANGLES FOLLOWING ANGULAR ADJUSTMENT =539 59 60.4999999869
SUM LOG SIN ODD ANGLES=-1.01900501943
SUM LOG SIN EVEN ANGLES=-1.01902086283
DIFFERENCE LOGSIN ODDS AND EVENS=158.434056274
DIFFERENCE FOR 1 SEC ANGLE1=25.6497870423
DIFFERENCE FOR 1 SEC ANGLE2=48.5571016387
DIFFERENCE FOR 1 SEC ANGLE3=96.1104123022
DIFFERENCE FOR 1 SEC ANGLE4=29.4696744324
DIFFERENCE FOR 1 SEC ANGLE5=21.1823232648
DIFFERENCE FOR 1 SEC ANGLE6=46.3120683711
SUM OF DIFFS FOR 1 SEC=267
SIDE CORRECTION =0.592761321233SECS

FINAL ADJUSTED VALUE FOR ANGLE1=39 22 52
FINAL ADJUSTED VALUE FOR ANGLE3=12 21 23
FINAL ADJUSTED VALUE FOR ANGLE5=44 49 38
FINAL ADJUSTED VALUE FOR ANGLE2=23 26 32
FINAL ADJUSTED VALUE FOR ANGLE4=35 32 40
FINAL ADJUSTED VALUE FOR ANGLE6=24 26 54
FINAL ADJUSTED VALUE FOR ANGLE7=117 10 35
FINAL ADJUSTED VALUE FOR ANGLE8=132 5 56
FINAL ADJUSTED VALUE FOR ANGLE9=110 43 29

SUM OF FINAL ANGLES =539 59 60.4999999869


100 PRINT "PROGRAM 4.12"
110 INIT
120 REM THIS PROGRAM PERFORMS AN EQUAL SHIFT ADJUSTMENT OF THE ANGLES OF
130 REM A CENTRE-POINT POLYGON
140 SET DEGREES
150 DIM A(20),D(20),M(20),S(20),Z(10),H(20),B(20),C(20)
```

```
160 PRINT "ENTER NUMBER OF PERIMETER SIDES OF POLYGON=";
170 INPUT N
180 N=N*3
190 FOR I=1 TO N
200 PRINT "ENTER DEGS,MINS,SECS OF ANGLE";I;"=";
210 INPUT D(I),M(I),S(I)
220 A(I)=D(I)+M(I)/60+S(I)/3600
230 PRINT @4;"OBSERVED ANGLE";I;"=";D(I);M(I);S(I)
240 NEXT I
250 FOR J=1 TO N/3
260 Z(J)=A(2*J-1)+A(2*J)+A(N-N/3+J)
270 PRINT @4;"SUM OF ANGLES OF TRIANGLE";J;"=";Z(J)
280 B(J)=Z(J)-180
290 O=ABS(B(J))
300 GOSUB 840
310 D(1)=D(1)*SGN(B(J))
320 PRINT @4;"MISCLOSURE TRIANGLE";J;"=";D(1);INT(M(1));S(1)
330 C(J)=-(B(J)/3)
340 A(2*J-1)=A(2*J-1)+C(J)
350 O=A(2*J-1)
360 GOSUB 840
370 PRINT @4;"FIRST ADJUSTED VALUE ANGLE";2*J-1;"=";D(1);INT(M(1));
380 PRINT @4;INT(S(1))
390 A(2*J)=A(2*J)+C(J)
400 O=A(2*J)
410 GOSUB 840
420 PRINT @4;"FIRST ADJUSTED VALUE ANGLE";2*J;"=";D(1);INT(M(1));
430 PRINT @4;INT(S(1))
440 A(N-N/3+J)=A(N-N/3+J)+C(J)
450 O=A(N-N/3+J)
460 GOSUB 840
470 PRINT @4;"FIRST ADJUSTED VALUE ANGLE";N-N/3+J;"=";
480 PRINT @4;D(1);INT(M(1));INT(S(1))
490 NEXT J
500 PRINT @4;
510 T1=0
520 FOR J=2*N/3+1 TO N
530 T1=T1+A(J)
540 NEXT J
550 PRINT @4;"SUM OF CENTRE POINT ANGLES=";T1
560 E=T1-360
570 O=ABS(E)
580 GOSUB 840
590 D(1)=D(1)*SGN(E)
600 PRINT @4;"MISCLOSURE CENTRE POINT ANGLES =";D(1);INT(M(1));S(1)
610 F=-E/(N/3)
620 FOR J=2*N/3+1 TO N
630 A(J)=A(J)+F
640 O=A(J)
650 GOSUB 840
660 PRINT @4;"SECOND ADJUSTED VALUES ANGLE";J;"=";D(1);INT(M(1));
670 PRINT @4;INT(S(1))
680 NEXT J
690 FOR J=1 TO N-N/3
700 A(J)=A(J)-F/2
710 O=A(J)
720 GOSUB 840
730 PRINT @4;"SECOND ADJUSTED VALUES ANGLE";J;"=";D(1);INT(M(1));
740 PRINT @4;INT(S(1))
750 NEXT J
760 T2=0
770 FOR I=1 TO N
780 T2=T2+A(I)
790 NEXT I
800 O=T2
810 GOSUB 840
820 PRINT @4;"SUM OF ANGLES FOLLOWING ANGULAR ADJUSTMENT =";D(1);
830 PRINT @4;INT(M(1));S(1)
835 GO TO 880
840 D(1)=INT(O)
```

```
850 M(1)=(O-D(1))*60
860 S(1)=(M(1)-INT(M(1)))*60+0.5
870 RETURN
880 P=0
890 FOR J=1 TO N-N/3 STEP 2
900 P=P+LGT(SIN(A(J)))
910 NEXT J
920 Q=0
930 FOR J=2 TO N-N/3 STEP 2
940 Q=Q+LGT(SIN(A(J)))
945 NEXT J
950 PRINT @4:"SUM LOG SIN ODD ANGLES=";P
960 PRINT @4:"SUM LOG SIN EVEN ANGLES=";Q
970 R=ABS((P-Q)*10^7)
980 PRINT @4:"DIFFERENCE LOGSIN ODDS AND EVENS=";R
990 U=0
1000 FOR J=1 TO N-N/3
1010 H(J)=21.055*(1/TAN(A(J)))
1020 PRINT @4:"DIFFERENCE FOR 1 SEC ANGLE";J;"=";H(J)
1030 U=U+H(J)
1040 NEXT J
1050 PRINT @4:"SUM OF DIFFS FOR 1 SEC=";INT(U)
1060 V=R/U/3600
1070 PRINT @4:"SIDE CORRECTION =";V*3600;"SECS"
1080 PRINT @4:
1090 FOR J=1 TO N-N/3 STEP 2
1100 A(J)=A(J)-V
1105 O=A(J)
1106 GOSUB 840
1110 PRINT @4:"FINAL ADJUSTED VALUE FOR ANGLE";J;"=";D(1);INT(M(1));
1120 PRINT @4:INT(S(1))
1130 NEXT J
1140 FOR J=2 TO N-N/3 STEP 2
1150 A(J)=A(J)+V
1160 O=A(J)
1170 GOSUB 840
1180 PRINT @4:"FINAL ADJUSTED VALUE FOR ANGLE";J;"=";D(1);INT(M(1));
1190 PRINT @4:INT(S(1))
1200 NEXT J
1210 FOR J=2*N/3+1 TO N
1215 O=A(J)
1216 GOSUB 840
1220 PRINT @4:"FINAL ADJUSTED VALUE FOR ANGLE";J;"=";D(1);INT(M(1));
1230 PRINT @4:INT(S(1))
1240 NEXT J
1250 T3=0
1260 FOR I=1 TO N-1
1270 T3=T3+A(I)
1280 NEXT I
1290 O=T3
1300 GOSUB 840
1310 PRINT @4:
1320 PRINT @4:"SUM OF FINAL ANGLES =";D(1);INT(M(1));S(1)
1330 STOP
1340 END
```

Example 4.12

Referring to Figure 4.2. Assume that the centre point polygon H(BCDEF) has been previously adjusted giving the following angles:

$$(18+23)=47°50'06''$$
$$4=93\ 59\ 53$$
$$13=38\ 10\ 01$$

Table 4.4 Solution to Example 4.12

1 Angle	2 Observed ° ' "	3 Adjustment to 360° "	4 Sum of opps ° ' "	5 Adjustment to opps "	6 First adjusted angles ° ' "	7 Log sines	8 Log sines	9 d	10 C"	11 Final angles ° ' "
4	93 59 53				93 59 53	-0.0010582		-		93 59 53
18	23 22 41		117 22 34	+1.7	23 22 42.7		-0.4014240	48.7	-0.9	23 22 41.8
23	24 27 25			-1.7	24 27 23.3	-0.3829977		46.3	+0.9	24 27 24.2
13	38 10 01 — 180 00 00		62 37 26		38 10 01 — 180 00 00.0		-0.2090432			38 10 01 — 180 00 00.0
14	48 21 19	+1.2		-0.9	48 21 19.3	-0.1265163		18.7	+0.9	48 21 20.2
15	69 01 16	+1.3	117 22 37.5	-0.9	69 01 16.4		-0.0297865	8.1	-0.9	69 01 15.5
16	26 09 33	+1.2	62 37 22.5	+0.9	26 09 35.1	-0.3556841		42.9	+0.9	26 09 36.0
3	36 27 47 — 179 59 55 — 359 59 55	+1.3		+0.9	36 27 49.2 — 180 00 00.0		-0.2259848	28.5	-0.9	36 27 48.3 — 180 00 00.0

$$-0.8662563$$
$$-0.8662385$$
$$-0.8662385$$

$$\frac{178}{193.2} = 0.9$$

d total = 193.2

The remaining observed angles in the braced quadrilateral FGBH are as follows:

	°	′	″		°	′	″
3.	36	27	47	14.	48	21	19
18.	23	22	42	15.	69	01	16
23.	24	27	26	16.	26	09	33

Determine adjusted values for the angles.

Solution (see Table 4.4):
Angles $(18+23)$, 4 and 13 may not be altered. The sum of the observed angles 18 and 23 must therefore be made equal to $47°50'06''$, that is:

	°	′	″		°	′	″
18.	23°	22′	42″	−1″	23°	22′	41″
23.	24	27	26	−1	24	27	25
	47	50	08	−2	47	50	06

Program 4.13 Triangulation adjustment example

```
OBSERVED ANGLES

ANGLE1=93 59 53
ANGLE2=23 22 42
ANGLE3=24 27 26
ANGLE4=38 10 1
ANGLE5=48 21 19
ANGLE6=69 1 16
ANGLE7=26 9 33
ANGLE8=36 27 47
FIRST ANGLE MISCLOSURE=5.555555554E-4

ANGLES FOLLOWING FIRST ADJUSTMENT
ANGLE 2=23.3780555556
ANGLE 3=24.4569444444
SUM OF ANGLES OF FIRST TRIANGLE=180

SUM OF ANGLES OF SECOND TRIANGLE=179.998611111
SECOND ANGLE MISCLOSURE=-0.00138888889069
ANGLES FOLLOWING 2ND ADJUSTMENT

ANGLE5=48.355625
ANGLE6=69.0214583333
ANGLE7=26.1595138889
ANGLE8=36.4634027778

THIRD ANGLE MISCLOSURE=0 0 4.00000000126
FOURTH ANGLE MISCLOSURE=0 0 3.99999999717
SUM OF ANGLES FOLLOWING 3RD AND 4TH ANGLE ADJUSTMENT=359 59 60
SUM OF LOG SIN ODD ANGLES=-0.866256311065
SUM OF LOG SIN EVEN ANGLES=-0.866238626469
DIFFERENCE IN SUMS OF LOG SINS=-1.768459621E-5
SUM OF DIFFS FOR ONE SECOND=193.152985046

ADJUSTED ANGLES

ANGLE1=93 59 53
ANGLE2=23 22 42
ANGLE3=24 27 24
```

```
ANGLE4=38 10 1
ANGLE5=48 21 20
ANGLE6=69 1 15
ANGLE7=26 9 36
ANGLE8=36 27 48
SUM OF FINAL ADJUSTED ANGLES =359 59 60


100 PRINT "PROGRAM 4.13"
110 INIT
120 REM THIS PROGRAM PERFORMS AN EQUAL SHIFT ADJUSTMENT OF A BRACED
130 REM QUADRILATERAL FORMING PART OF A PREVIOUSLY ADJUSTED FIGURE
140 REM (EXAMPLE 4.12)
150 SET DEGREES
160 DIM A(8),T(8),D(10),M(10),S(10),Z(5),J(5),L(5)
170 PRINT @4:"OBSERVED ANGLES"
180 PRINT @4:
190 Z(1)=0
200 FOR I=1 TO 8
210 READ D(I),M(I),S(I)
220 DATA 93,59,53,23,22,42,24,27,26,38,10,1,48,21,19,69,1,16,26,9,33
230 DATA 36,27,47
240 PRINT @4:"ANGLE";I;"=";D(I);M(I);S(I)
250 A(I)=D(I)+M(I)/60+S(I)/3600
260 NEXT I
270 READ D(9),M(9),S(9)
280 DATA 47,50,6
290 X1=D(9)+M(9)/60+S(9)/3600
300 X2=A(2)+A(3)
310 X3=X2-X1
320 PRINT @4:"FIRST ANGLE MISCLOSURE=";X3
330 PRINT @4:
340 A(2)=A(2)-X3/2
350 A(3)=A(3)-X3/2
360 PRINT @4:"ANGLES FOLLOWING FIRST ADJUSTMENT"
370 PRINT @4:"ANGLE 2=";A(2)
380 PRINT @4:"ANGLE 3=";A(3)
390 Z1=A(1)+A(2)+A(3)+A(4)
400 PRINT @4:"SUM OF ANGLES OF FIRST TRIANGLE=";Z1
410 PRINT @4:
420 Z2=A(5)+A(6)+A(7)+A(8)
430 PRINT @4:"SUM OF ANGLES OF SECOND TRIANGLE=";Z2
440 Z3=Z2-180
450 PRINT @4:"SECOND ANGLE MISCLOSURE=";Z3
460 PRINT @4:"ANGLES FOLLOWING 2ND ADJUSTMENT"
470 PRINT @4:
480 FOR I=5 TO 8
490 A(I)=A(I)-Z3/4
500 PRINT @4:"ANGLE";I;"=";A(I)
510 NEXT I
520 E=A(1)+A(2)
530 F=A(6)+A(5)
540 G=A(3)+A(4)
550 H=A(7)+A(8)
560 J(1)=E-F
570 O=ABS(J(1))
580 GOSUB 1190
590 PRINT @4:
600 PRINT @4:"THIRD ANGLE MISCLOSURE=";D(1);INT(M(1));S(1)
610 K=-(J(1)/4)
620 A(2)=A(2)+2*K
630 A(5)=A(5)-K
640 A(6)=A(6)-K
650 J(2)=A(1)+A(2)-(A(5)+A(6))
660 L(1)=G-H
670 O=ABS(L(1))
680 GOSUB 1190
690 D(1)=D(1)*SGN(L(1))
700 PRINT @4:"FOURTH ANGLE MISCLOSURE=";D(1);INT(M(1));S(1)
710 N=-(L(1)/4)
720 A(3)=A(3)+2*N
```

```
730 A(7)=A(7)-N
740 A(8)=A(8)-N
750 L(2)=A(3)+A(4)-(A(7)+A(8))
760 Z(2)=A(1)+A(2)+A(3)+A(4)+A(5)+A(6)+A(7)+A(8)
770 O=Z(2)
780 GOSUB 1190
790 PRI @4:"SUM OF ANGLES FOLLOWING 3RD AND 4TH ANGLE ADJUSTMENT=";D(1);
800 PRINT @4:INT(M(1));INT(S(1))
810 P=LGT(SIN(A(1)))+LGT(SIN(A(3)))+LGT(SIN(A(5)))+LGT(SIN(A(7)))
820 PRINT @4:"SUM OF LOG SIN ODD ANGLES=";P
830 Q=LGT(SIN(A(2)))+LGT(SIN(A(4)))+LGT(SIN(A(6)))+LGT(SIN(A(8)))
840 PRINT @4:"SUM OF LOG SIN EVEN ANGLES=";Q
850 R=P-Q
860 PRINT @4:"DIFFERENCE IN SUMS OF LOG SINS=";R
870 R=R*10^7
880 U=0
890 FOR I=2 TO 3
900 T(I)=21.055*(1/TAN(A(I)))
910 U=U+T(I)
920 NEXT I
930 FOR I=5 TO 8
940 T(I)=21.055*(1/TAN(A(I)))
950 U=U+T(I)
960 NEXT I
970 PRINT @4:"SUM OF DIFFS FOR ONE SECOND=";U
980 V=ABS(R/U/3600)
990 PRINT @4:
1000 A(2)=A(2)-V
1010 A(3)=A(3)+V
1020 A(5)=A(5)+V
1030 A(6)=A(6)-V
1040 A(7)=A(7)+V
1050 A(8)=A(8)-V
1060 PRINT @4:"ADJUSTED ANGLES"
1070 PRINT @4:
1080 FOR I=1 TO 8
1090 D(I)=INT(A(I))
1100 M(I)=(A(I)-D(I))*60
1110 S(I)=(M(I)-INT(M(I)))*60+0.5
1120 PRINT @4:"ANGLE";I;"=";D(I);INT(M(I));INT(S(I))
1130 NEXT I
1140 Z(2)=A(1)+A(2)+A(3)+A(4)+A(5)+A(6)+A(7)+A(8)
1150 Z(2)=0
1160 GOSUB 1190
1170 PRINT @4:"SUM OF FINAL ADJUSTED ANGLES =";D(1);INT(M(1));INT(S(1))
1180 GO TO 1230
1190 D(1)=INT(O)
1200 M(1)=(O-D(1))*60
1210 S(1)=(M(1)-INT(M(1)))*60+0.5
1220 RETURN
1230 STOP
1240 END
```

4.1.3 Eccentric station reduction

A point which may be intersected (see Section 4.3) but which cannot be occupied (the 'parent' station) may still be included in a triangulation network by use of an *eccentric* station located as close to the parent station as possible. Observations are taken from the eccentric station to distant control points and to the parent station and the horizontal distance to the parent station is determined. The corrections necessary to convert the observed angles at the eccentric station into corresponding angles at the parent station may be derived as follows.

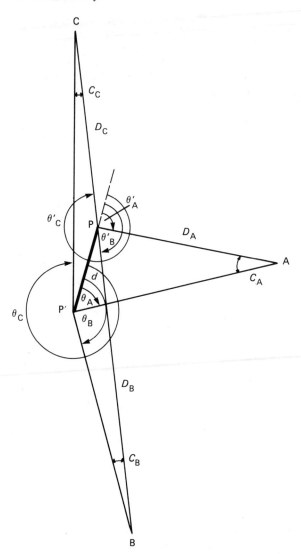

Figure 4.5 Eccentric reduction

Referring to Figure 4.5, P is the parent station distant d from P', the eccentric. θ_A, θ_B and θ_C are the angles measured at P' clockwise from P to each of the distant control points A, B and C and C_A, C_B and C_C are the small angles at each of these points subtended by the distance d.

Then:

$$\sin C_A = \frac{d\sin\theta_A}{D_A}, \quad \sin C_B = \frac{d\sin\theta_B}{D_B} \quad \text{etc} \qquad (4.12a)$$

or in seconds of arc:

$$C_A'' = \frac{d\sin\theta_A}{D_A\sin 1'''}, \quad C_B'' = \frac{d\sin\theta_B}{D_B\sin 1''} \quad \text{etc} \qquad (4.12b)$$

Note that the sign of the correction is $+$ if $\theta < 180°$ and is $-$ if $\theta > 180°$.

The angles adjusted 'to centre' (to the parent station) paying due regard to the signs of the corrections, are:

$$\theta_A' = \theta_A + C_A$$
$$\theta_B' = \theta_B + C_B \quad \text{etc}$$

Example 4.13
Referring to Figure 4.5 observations taken from an eccentric station P' to A, B and C were as follows:

Station	Mean horizontal circle reading	Distance (m)
P	10°10'10"	1.926
A	65 55 50	643
B	150 22 10	1226
C	345 42 30	1579

Determine the reduced angles at P.

Solution (see Table 4.5):

$$\frac{d}{\sin 1''} = \text{constant} = 397\,266.35$$

The angles required are determined thus:

Table 4.5 Solution to Example 4.13

Station	Observation			θ			sin θ	D	Correction	Corrected observation		
	°	′	″	°	′	″				°	′	″
P	10	10	10	55	45	40	+0.826699	643	+8631″	66	04	21
A	65	55	50	140	12	00	+0.640110	1226	+3′27″	150	25	37
B	150	22	10	335	32	20	−0.414076	1579	−1′44″	345	40	46
C	345	42	30									

Then angles reduced to P(θ′) are: A = 55°54′11″, B = 140°15′27″. C = 335°30′36″.

Program 4.14 Eccentric station adjustment

```
ANGLE1=55 45 40
CORRECTION FOR DATUM STATION1=510.761552106SECS
ANGLE2=140 12 0
CORRECTION FOR DATUM STATION2=207.417674974SECS
ANGLE3=335 32 20
CORRECTION FOR DATUM STATION3=-104.17877603SECS

ANGLES REDUCED TO PARENT STATION
ANGLE1=55 54 11
ANGLE2=140 15 27
ANGLE3=335 30 36
```

```
100 PRINT "PROGRAM 4.14"
110 REM THIS PROGRAM REDUCES HORIZONTAL ANGLE OBSERVATIONS FROM AN
120 REM ECCENTRIC STATION TO EQUIVALENT OBSERVATIONS FROM
130 REM THE INACCESSIBLE PARENT STATION
140 INIT
150 SET DEGREES
160 PRINT "ENTER REDUCED DISTANCE ECCENTRIC TO PARENT STATION=";
170 INPUT B
180 PRINT "ENTER NUMBER OF DATUM STATIONS OBSERVED=";
190 INPUT N
200 DIM A(N),D(N),M(N),S(N),C(N),L(N),P(N)
210 FOR I=1 TO N
220 PRINT "ENTER OBSERVED CLOCKWISE ANGLE FROM ECC STATION TO CONTROL"
230 PRINT "POINT";I;"=";
240 INPUT D(I),M(I),S(I)
250 PRINT @4:"ANGLE";I;"=";D(I);M(I);S(I)
260 A(I)=D(I)+M(I)/60+S(I)/3600
270 PRINT "ENTER DISTANCE ECCENTRIC/PARENT STATION TO CONTROL POINT";I;
280 INPUT L(I)
290 C(I)=B*SIN(A(I))/(L(I)*(1/206265))
300 PRINT @4:"CORRECTION FOR DATUM STATION";I;"=";C(I);"SECS"
310 C(I)=C(I)/3600
320 P(I)=A(I)+C(I)
330 NEXT I
340 PRINT @4:
350 PRINT @4:"ANGLES REDUCED TO PARENT STATION"
360 FOR I=1 TO N
370 D(I)=INT(P(I))
380 M(I)=(P(I)-D(I))*60
390 S(I)=(M(I)-INT(M(I)))*60+0.5
400 PRINT @4:"ANGLE";I;"=";D(I);INT(M(I));INT(S(I))
410 NEXT I
420 STOP
430 END
```

4.1.4 Trigonometric heighting

The difference in elevation between two points can be determined by observing the vertical angle of the line from one point to the other and then computing the difference in elevation from a knowledge of the horizontal distance between the two points (Figure 4.6). The horizontal distance may be derived from the known coordinates of the terminals or from direct measurement of the slope distance.

In order to eliminate or reduce the effects of earth curvature and atmospheric refraction, vertical angle observations are usually made at both ends of the line as near simultaneously as possible (Figure 4.7). Such *reciprocal* observations may be used to derive the height

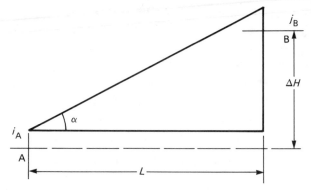

Figure 4.6 Trigonometric heighting: basic principles

(H_B) of a point B from a point A of known height (H_A) using the following relationship:

$$H_B = H_A + L \tan \tfrac{1}{2}(\alpha + \beta) + \tfrac{1}{2}(i_A + j_A) - \tfrac{1}{2}(i_B + j_B) \qquad (4.13)$$

where L = horizontal distance between points A and B, α, β = observed vertical angles at A and B respectively, i_A, i_B = height of instrument at A and B respectively, j_A, j_B = height of signal at A and B respectively.

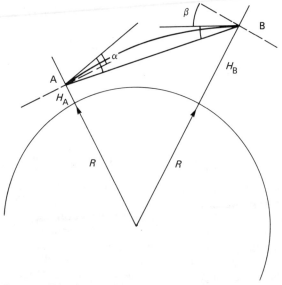

Figure 4.7 Trigonometric heighting: effect of atmospheric refraction

When a line has been observed in one direction only, a correction for earth curvature and atmospheric refraction must be applied, and:

$$H_B = H_A + L\tan\alpha + \frac{(1-2k)}{2R}L^2 + i_A - j_B \tag{4.14}$$

where α = vertical angle measured at the instrument station (positive for elevation, negative for depression), k = coefficient of refraction, and R = radius of the earth.

A value for k may be deduced from simultaneously observed reciprocal vertical angles. If the instrument and signal heights are converted to angular corrections, modified vertical angles may be deduced by applying such corrections to the observed angles. The correction in seconds of arc for each angle may be determined from:

$$c'' = \frac{i_A - j_B}{L}\operatorname{cosec}1'' \tag{4.15a}$$

and:

$$k = 0.5 - \tfrac{1}{2}(\beta' - \alpha')\sin 1''\frac{R}{L} \tag{4.15b}$$

where α', β' = the absolute numerical values of the modified vertical angles in seconds of arc = $\alpha + c$, $\beta + c$ respectively.

Alternatively, when observations have been made between points of known height

$$ks = 0.5 - \frac{R}{L^2}(H_B - H_A - L\tan\alpha) \tag{4.16}$$

Example 4.14(a)
Determine the altitude of B using the following data and assuming that observations at A and B were truly reciprocal:

At A:
    Vertical angle to B = $+1°10'44''$
    Height of instrument at A = 1.585 m
    Height of signal at A = 2.770 m

At B:
    Vertical angle to A = $-1°14'07''$
    Height of instrument at B = 1.533 m
    Height of signal at B = 3.140 m
    Altitude of A = 506.24 m   Distance AB = 9430 m

Solution:
From Equation (4.13):

$$H_B = 506.24 + 9430 \tan \tfrac{1}{2}(2°24'51'') + \tfrac{1}{2}(4.355) - \tfrac{1}{2}(4.673)$$
$$= 506.24 + 198.697 + 2.1775 - 2.3365$$
$$= 704.78 \text{ m}$$

Example 4.14(b)
Using the data in Example 4.14(a) determine the value of the coefficient of refraction.

Solution:
Using Equation (4.15):

Modified vertical angles:

$$\text{From A to B} \quad c'' = \left(\frac{1.585 - 3.140}{9430}\right) 206\,265 = -34''$$

$$\text{From B to A} \quad c'' = \left(\frac{1.533 - 2.770}{9430}\right) 206\,265 = -27''$$

Modified vertical angles
  A to B = 1°10'44'' − 34'' = 1°10'10''
  B to A = −1°14'07'' − 27'' = −1°14'34''

$$k = 0.5 + \frac{\tfrac{1}{2}(-264'') \times 6378}{9.43 \times 20\,625} = 0.067$$

Example 4.14(c)
Determine the altitude of B if the observations at A and B were not truly reciprocal. Assume the coefficient of refraction = 0.07 and the radius of the earth = 6378 km.

Solution:
From Equation (4.14):

From *A*:

$$\Delta H_{AB} = +9430 \tan 1°10'44'' + \left[\frac{1 - (2 \times 0.07)}{2 \times (6378 \times 10^3)}\right] 9430^2$$
$$+ 1.585 - 3.140$$
$$= +194.054 + 5.995 + 1.585 - 3.140$$
$$= +198.494 \text{ m}$$

From *B*:

$$\Delta H_{BA} = -9430 \tan 1°14'07'' + \left[\frac{1 - (2 \times 0.07)}{2 \times (6378 \times 10^3)}\right] 9430^2$$

$$+1.533 - 2.770$$
$$= -203.339 + 5.995 + 1.533 - 2.770$$
$$= -198.581$$

Mean $\Delta H = 198.54$

$H_B = 506.24 + 198.54 = 704.78$ m

Program 4.15 Trigonometric heighting

```
EXAMPLE 4.14(A)

ALT OF A=506.24  HT INST A=1.585  HT SIG A=2.77
HT INST B=1.533  HT SIG B=3.14
V.A. AT A=1 10 44  V.A. AT B=-1 14 7
DISTANCE A TO B=9430
DIFFERENCE IN HEIGHT AB=198.537662943
ALTITUDE OF B=704.777662943


EXAMPLE 4.14(B)

ALTITUDE OF A=506.24  ALTITUDE OF B=704.78HEIGHT INST AT A=1.585
HEIGHT SIGNAL AT A=2.77
HEIGHT INST AT B=1.533  HEIGHT SIGNAL AT B=3.14
V.A AT A=1 10 44  V.A AT B=-1 14 7
DISTANCE A TO B =9430
COEFFICIENT OF REFRACTION =0.0670510793645


EXAMPLE 4.14(C)

ALT OF A=506.24  HT INST A=1.585  HT SIG A=2.77
HT INST B=1.533  HT SIG B=3.14
V.A AT A=1 10 44  V.A AT B=-1 14 7
DISTANCE A TO B=9430
DIFFERENCE IN HT A TO B=198.494530982
DIFFERENCE IN HT B TO A=-198.580891174
MEAN DIFFERNCE IN HT=198.537711078
ALTITUDE OF B=704.777711078


100 PRINT "PROGRAM 4.15"
110 REM THIS PROGRAM REFERS TO EXAMPLES 4.14 (A,B AND C). THE FIRST
120 REM SECTION DETERMINES THE COEFFICIENT OF REFRACTION GIVEN
130 REM RECIPROCAL VERTICAL ANGLE OBSERVATIONS BETWEEN TWO POINTS OF
140 REM KNOWN ALTITUDE. THE SECOND SECTION DETERMINES THE HEIGHT
150 REM DIFFERENCE BETWEEN TWO POINTS GIVEN RECIPROCAL VERTICAL
160 REM ANGLES AT EACH POINT.THE THIRD SECTION DETERMINES THE HEIGHT
170 REM DIFFERENCE BETWEEN TWO POINTS GIVEN NON-RECIPROCAL OBSERVATIONS
180 INIT
190 SET DEGREES
200 R=6378000
210 PRINT "INDICATE WHICH OF THE FOLLOWING OPTIONS YOU REQUIRE BY"
220 PRINT "PRESSING THE APPROPRIATE KEY"
230 PRINT "COMPUTATION COMPLETED. PRESS O"
240 PRINT "SECTION1.DETERMINATION OF COEFFICIENT OF REFRACTION. PRESS 1"
250 PRINT "SECTION2.HEIGHT DIFFERENCE USING RECIPROCAL OBSERVATIONS."
260 PRINT "PRESS 2"
270 PRINT "SECTION3.HEIGHT DIFFERENCE USING NON-RECIP OBSERVATIONS."
280 PRINT "PRESS 3"
290 PRINT "ENTER OPTION NUMBER";
300 INPUT Q1
310 IF Q1=0 THEN 1350
320 IF Q1=2 THEN 690
330 IF Q1=3 THEN 1010
340 PRINT "SECTION 1"
350 PRINT "ENTER ALTITUDE A, ALTITUDE B";
```

```
360 INPUT H1,H2
370 PRINT "ENTER DEGREES,MINS,SECS OF VERTICAL ANGLE AT A";
380 INPUT D1,M1,S1
390 V1=(ABS(D1)+M1/60+S1/3600)*SGN(D1)
400 PRINT "ENTER DEGREES,MINS,SECS OF VERTICAL ANGLE AT B";
410 INPUT D2,M2,S2
420 V2=(ABS(D2)+M2/60+S2/3600)*SGN(D2)
430 PRINT "ENTER HEIGHT OF INST AT A, HEIGHT OF SIGNAL AT A";
440 INPUT I1,J1
450 PRINT "ENTER HEIGHT OF INST AT B, HEIGHT OF SIGNAL AT B";
460 INPUT I2,J2
470 PRINT "ENTER DISTANCE A TO B";
480 INPUT L
490 C1=(I1-J2)/L*206265/3600
500 C2=(I2-J1)/L*206265/3600
510 V1=V1+C1
520 V2=V2+C2
530 K=0.5-(ABS(V2)-ABS(V1))/2*3600/206265*(R/L)
540 PRINT @4:
550 PRINT @4:
560 PRINT @4:"EXAMPLE 4.14(B)"
570 PRINT @4:
580 PRINT @4:"ALTITUDE OF A=";H1;"  ";"ALTITUDE OF B=";H2;
590 PRINT @4:"HEIGHT INST AT A=";I1;"  ";"HEIGHT SIGNAL AT A=";J1
600 PRINT @4:"HEIGHT INST AT B=";I2;"   ";"HEIGHT SIGNAL AT B=";J2
610 PRINT @4:"V.A AT A=";D1;M1;S1;"   ";"V.A AT B=";D2;M2;S2
620 PRINT @4:"DISTANCE A TO B =";L
630 PRINT @4:"COEFFICIENT OF REFRACTION =";K
640 PRINT "DO YOU WISH TO REPEAT THIS OPTION?"
650 PRINT "IF YES PRESS 1; IF NO PRESS 0"
660 INPUT Q2
670 IF Q2=0 THEN 210
680 IF Q2=1 THEN 350
690 PRINT "SECTION 2"
700 PRINT "ENTER ALT OF A,HT INST A,HT SIG A=";
710 INPUT H1,I1,J1
720 PRINT "ENTER DEGREES,MINS,SECS OF VERTICAL ANGLE AT A="
730 INPUT D1,M1,S1
740 PRINT "ENTER HEIGHT OF INST B,HEIGHT OF SIG B=";
750 INPUT I2,J2
760 PRINT "ENTER DEGREES,MINS,SECS OF VERTICAL ANGLE AT B=";
770 INPUT D2,M2,S2
780 PRINT "ENTER DISTANCE A TO B=";
790 INPUT L
800 V1=(ABS(D1)+M1/60+S1/3600)*SGN(D1)
810 V2=(ABS(D2)+M2/60+S2/3600)*SGN(D2)
820 H3=L*TAN(ABS(V1)+ABS(V2)/2)+(I1+J1)/2-(I2+J2)/2
830 V3=(ABS(V1)+ABS(V2))/2
840 H3=L*TAN(V3)+(I1+J1)/2-(I2+J2)/2
850 H2=H1+H3*SGN(D1)
860 PRINT @4:
870 PRINT @4:
880 PRINT @4:"EXAMPLE 4.14(A)"
890 PRINT @4:
900 PRINT @4:"ALT OF A=";H1;"  ";"HT INST A=";I1;"  ";"HT SIG A=";J1
910 PRINT @4:"HT INST B=";I2;"  ";"HT SIG B=";J2
920 PRINT @4:"V.A. AT A=";D1;M1;S1;"  ";"V.A. AT B=";D2;M2;S2
930 PRINT @4:"DISTANCE A TO B=";L
940 PRINT @4:"DIFFERENCE IN HEIGHT AB=";H3
950 PRINT @4:"ALTITUDE OF B=";H2
960 PRINT "DO YOU WISH TO REPEAT THIS COMPUTATION?"
970 PRINT "IF YES PRESS 1; IF NO PRESS 0"
980 INPUT Q3
990 IF Q3=0 THEN 210
1000 IF Q3=1 THEN 690
1010 PRINT "SECTION 3"
1020 PRINT "ENTER ALTITUDE A,HEIGHT OF INST A,HEIGHT OF SIG A=";
1030 INPUT H1,I1,J1
1040 PRINT "ENTER HEIGHT OF INST B,HEIGHT OF SIG B=";
1050 INPUT I2,J2
1060 PRINT "ENTER DEGREES,MINS,SECS OF VERTICAL ANGLE AT A=";
1070 INPUT D1,M1,S1
```

```
1080 PRINT "ENTER DEGREES,MINS,SECS OF VERTICAL ANGLE AT B=";
1090 INPUT D2,M2,S2
1100 PRINT "ENTER COEFFCIENT OF REFRACTION, DISTANCE A TO B=";
1110 INPUT K,L
1120 V1=(ABS(D1)+M1/60+S1/3600)*SGN(D1)
1130 V2=(ABS(D2)+M2/60+S2/3600)*SGN(D2)
1140 H3=L*TAN(V1)+(1-2*K)/(2*R)*L^2+I1-J2
1150 H4=L*TAN(V2)+(1-2*K)/(2*R)*L^2+I2-J1
1160 H5=(ABS(H3)+ABS(H4))/2
1170 H2=H1+H5*SGN(D1)
1180 PRINT @4;
1190 PRINT @4;
1200 PRINT @4;"EXAMPLE 4.14(C)"
1210 PRINT @4;
1220 PRINT @4;"ALT OF A=";H1;"   ";"HT INST A=";I1;"   ";"HT SIG A=";J1
1230 PRINT @4;"HT INST B=";I2;"   ";"HT SIG B=";J2
1240 PRINT @4;"V.A AT A=";D1;M1;S1;"   ";"V.A AT B=";D2;M2;S2
1250 PRINT @4;"DISTANCE A TO B=";L
1260 PRINT @4;"DIFFERENCE IN HT A TO B=";H3
1270 PRINT @4;"DIFFERENCE IN HT B TO A=";H4
1280 PRINT @4;"MEAN DIFFERNCE IN HT=";H5
1290 PRINT @4;"ALTITUDE OF B=";H2
1300 PRINT "DO YOU WISH TO REPEAT THIS COMPUTATION?"
1310 PRINT "IF YES PRESS 1; IF NO PRESS 0";
1320 INPUT Q4
1330 IF Q4=0 THEN 210
1340 IF Q4=1 THEN 1010
1350 STOP
```

4.2 Trilateration

If, instead of measuring the angles of a triangulation network, the lengths of the sides of the component figures are measured, the position of control points may be determined using *trilateration*.

Given the coordinates of two points A and B (Figure 4.8) and the reduced horizontal distances from these points to a third point C, the angles of the triangle may be deduced as follows:

$$\sin A/2 = \left[\frac{(p-b)(p-c)}{bc} \right]^{1/2} \qquad (4.17a)$$

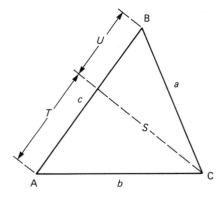

Figure 4.8 Trilateration

$$\cos A/2 = \left[\frac{p(p-a)}{bc} \right]^{1/2} \tag{4.17b}$$

$$\tan A/2 = \left[\frac{(p-b)(p-c)}{p(p-a)} \right]^{1/2} \tag{4.17c}$$

where p = the semiperimeter = $(a+b+c)/2$.

The derived angles may be used to determine the bearings of the network lines and hence the coordinates of network points.

Alternatively the position of C may be determined directly from the coordinates of A and B, and the reduced side lengths a, b and c. Assuming a clockwise configuration of A, B and C then:

$$E_C = E_A + VS + WT = E_B + VS - WU \tag{4.18a}$$

$$N_C = N_A + VT - WS = N_B - VU - WS \tag{4.18b}$$

where:

$$V = \frac{N_B - N_A}{c}, \quad W = \frac{E_B - E_A}{c}$$

$$T = \frac{b^2 + c^2 - a^2}{2c}, \quad U = c - T$$

$$S = (b^2 - T^2)^{1/2} = (a^2 - U^2)^{1/2}$$

Example 4.15 (Figure 4.9)
Distances from A and B to a point C are measured with the following results: BC = 754.46 m, AC = 2962.43 m.

Assuming C to be to the right of AB and given the following coordinates, determine (*a*) the angles of the triangle ABC, and (*b*) the coordinates of C.

	E	(m)	N
B	28905.74		82282.61
A	29974.47		79634.85

Solution:
(*a*) $\Delta E_{AB} - 1068.73$ $\Delta N_{AB} + 2647.76$
Distance AB = $(1068.73^2 + 2647.76^2)^{1/2}$
= 2855.3138

$$p = \frac{2962.43 + 754.46 + 2855.314}{2} = 3286.1019$$

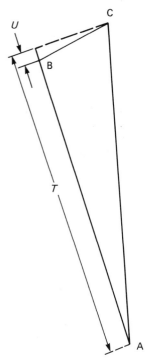

Figure 4.9 Trilateration example

Demonstrating the use of each formula:
From Equation (4.17a):

$$\sin B/2 = \left[\frac{(3286.102 - 754.46)(3286.102 - 2855.314)}{754.46 \times 2855.314} \right]^{1/2}$$

$$B = 90°43'03.6''$$

From Equation (4.17b):

$$\cos A/2 = \left[\frac{3286.102\,(3286.102 - 754.46)}{2962.43 \times 2855.314} \right]^{1/2}$$

$$A = 14°45'11.6''$$

From Equation (4.17c):

$$\tan C/2 = \left[\frac{(3286.102 - 2962.43)(3286.102 - 754.46)}{3286.102\,(3286.102 - 2855.314)} \right]^{1/2}$$

$$C = 74°31'44.9''$$

and:

$$A + B + C = 180°00'00''$$

(b) $$V = \frac{82282.61 - 79634.85}{2855.314} = +0.927\,309\,57$$

$$W = \frac{28905.74 - 29974.47}{2855.314} = -0.374\,295\,086$$

$$T = \frac{(2962.43)^2 + (2855.314)^2 - (754.46)^2}{2 \times 2855.314} = 2864.76$$

$$U = 2855.314 - 2864.764 = -9.450$$

$$S = [(754.46)^2 - (9.450)^2]^{1/2} = 754.40$$

Also:

$$S = [(2962.43)^2 - (2864.764)^2]^{1/2} = 754.40 \text{ (check)}$$

Then:

$$\begin{aligned} E_C &= 29974.47 + (0.927\,309\,57 \times 754.40) \\ &\quad + (-0.374\,295\,086 \times 2864.764) \\ &= 29601.77 \end{aligned}$$

Also:

$$\begin{aligned} E_C &= 28905.74 + (0.927\,309\,57 \times 754.40) \\ &\quad - (-0.374\,295\,086 \times -9.45) \\ &= 29601.77 \quad \text{(check)} \end{aligned}$$

And:

$$\begin{aligned} N_C &= 79634.85 + (0.927\,309\,57 \times 2864.76) \\ &\quad - (-0.374\,295\,086 \times 754.40) \\ &= 82573.74 \end{aligned}$$

Also:

$$\begin{aligned} N_C &= 82282.61 - (0.927\,309\,57 \times -9.45) \\ &\quad - (-0.374\,295\,086 \times 754.40) \\ &= 82573.74 \text{ (check)} \end{aligned}$$

Program 4.16 Trilateration

```
AC=2962.43BC=754.46

EASTINGS OF C=29601.7661328   NORTHINGS OF C=82573.741293

CHECK

EASTINGS OF C=29601.7661328   NORTHINGS OF C=82573.741293


100 PRINT "PROGRAM 4.16"
110 REM THIS PROGRAM DETERMINES THE RECTANGULAR COORDINATES OF A POINT
120 REM C GIVEN THE COORDINATES OF POINTS A AND B (C BEING TO THE RIGHT
130 REM OF AB) AND THE REDUCED HORIZONTAL DISTANCES AC AND BC.
140 DIM E(2),N(2)
150 PRINT "ENTER EASTINGS, NORTHINGS OF A=";
160 INPUT E(1),N(1)
170 PRINT "ENTER EASTINGS, NORTHINGS OF B=";
180 INPUT E(2),N(2)
190 PRINT "ENTER DISTANCE AB=";
200 INPUT D
210 PRINT "ENTER DISTANCES AC,BC=";
220 INPUT L1,L2
230 P=(L1+L2+D)/2
240 V=(N(2)-N(1))/D
250 W=(E(2)-E(1))/D
260 T=(L1^2+D^2-L2^2)/(2*D)
270 U=D-T
280 S=SQR(L1^2-T^2)
290 E3=E(1)+V*S+W*T
300 N3=N(1)+V*T-W*S
310 E4=E(2)+V*S-W*U
320 N4=N(2)-V*U-W*S
330 PRINT @4:"AC=";L1;"BC=";L2
340 PRINT @4:
350 PRINT @4:"EASTINGS OF C=";E3;"   ";"NORTHINGS OF C=";N3
360 PRINT @4:
370 PRINT @4:"CHECK"
380 PRINT @4:
390 PRINT @4:"EASTINGS OF C=";E4;"   ";"NORTHINGS OF C=";N4
400 STOP
410 END
```

4.3 Intersection

The coordination of an inaccessible station using the forward bearings from two fixed datum points was discussed in Section 3.3.2. If the forward bearings from additional datum points are also available all bearings may be used to coordinate the unknown station by a semigraphic intersecti on.

If the point is coordinated using the bearings from successive pairs of datum points, different sets of coordinates for the intersected point will result due to the effects of random errors in the observations from which the bearings are derived. Figure 4.10(*a*) illustrates the situation using three datum points A, B and C. The three intersections form the vertices of the triangle of error P_1, P_2, P_3 within which the most likely position of the unknown point (P) must be located.

To construct the triangle of error graphically the coordinates of a

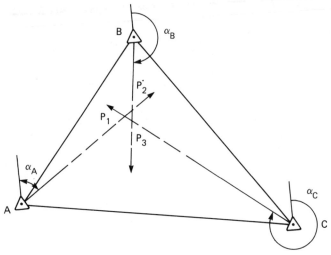

Figure 4.10(a)–(e) Semigraph intersection. (a) above

trial point are first determined using the forward bearings from two of the datum points and Equation (3.5) or (3.6). In Figure 4.10(*b*) the coordinates of P_1 have been determined from A and C. These trial coordinates then form the axes of the error figure diagram (Figure 4.10(*c*)) drawn at a suitably large scale and on which the directions of the lines P_1A and P_1C are first plotted.

Unless the bearing of the ray from B is 90°, 180°, 270° or 360°, it will naturally cut the P_1 axes in two positions (I_B and $I_{B'}$). These positions may be determined as follows:

$$\left. \begin{aligned} E_{IB} &= E_B - E_B I_B = E_B + (\Delta N_{BP1} \tan \alpha_B) \\ N_{IB} &= N_{P1} \end{aligned} \right\} \tag{4.19a}$$

and:

$$\left. \begin{aligned} N_{IB'} &= N_B - N_B I_{B'} = N_B + (\Delta E_{BP1} \cot \alpha_B) \\ E_{IB'} &= E_{P1} \end{aligned} \right\} \tag{4.19b}$$

The ray from B may be plotted using its known bearing and one of the above two cutting points. The point at which the ray makes the less acute angle with the relevant axis is therefore chosen (I_B in Figure 4.10(*c*)). To determine which of the two alternative points is most suitable the partial coordinates between the control point and the trial point ($\Delta E_{BP'}$ and $\Delta N_{BP'}$) are inspected. The largest of the two partial coordinates used in conjunction with the tan/cot function of

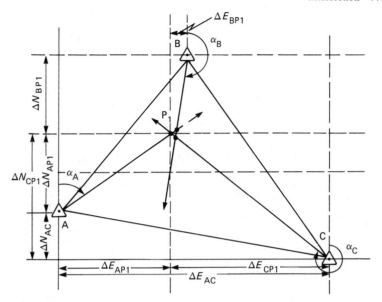

Figure 4.10(b)

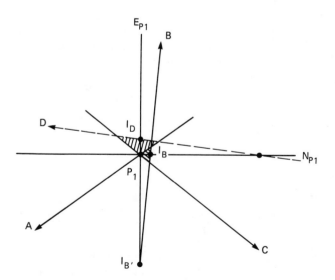

Figure 4.10(c)

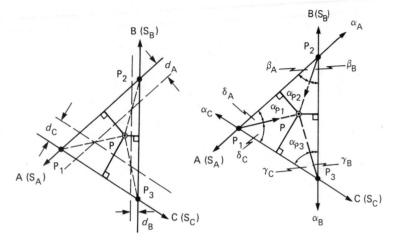

Figure 4.10(d) (left) and (e) (right)

the inward bearing (from **B**) that is less than unity will give the best cut.

The triangle of error is formed by plotting the position of the cut point (I_B) and drawing a line through this point at a bearing equal to the observed bearing from the relevant datum point (**B**).

Assuming that the forward bearings are of equal reliability, the magnitude of the displacement of **P** from the sides of the triangle of error will be proportional to the corresponding lengths of each fixing ray. The location of **P** may then be determined in one of the following ways:

1. (Figure 4.10(*d*)). By constructing a line parallel to each ray and at a distance from it proportional to the length of the observed ray, that is:

$$\frac{d_A}{S_A} = \frac{d_B}{S_B} = \frac{d_C}{S_C}$$

   Lines are then drawn connecting the intersection of pairs of rays with the intersection of the corresponding pair of parallels. The trisection thus formed locates **P**.

2. (Figure 4.10(*e*)). The angles of the error triangle may be deduced from the bearings of the fixing rays and divided, the ratio of the two resulting parts of each angle being made equal to the ratio of the lengths of the two corresponding rays, that is:

$$\frac{\delta_A}{\delta_C}=\frac{S_A}{S_C}, \quad \frac{\beta_A}{\beta_B}=\frac{S_A}{S_B}, \quad \frac{\gamma_B}{\gamma_C}=\frac{S_B}{S_C}$$

The intersection of the three lines so produced locates P.

3. The bearings of the lines forming the trisection and the coordinates of P_1, P_2 and P_3 may be used, that is:

$$\alpha_{P1}=\alpha_A+\delta_A=(\alpha_C-180)-\delta_C$$
$$\alpha_{P2}=\alpha_B+\beta_B=(\alpha_A+180)-\beta_A$$
$$\alpha_{P3}=\alpha_C+\gamma_C=(\alpha_B+180)-\gamma_B$$

P may then be coordinated using Equation (3.5) or (3.6).

Example 4.16
Determine the coordinates of P using the following data:

Station	E (m)	N	Bearing to P
A	428 905.74	182282.61	248°44′22″
B	429 974.47	179 634.85	323 16 02
C	423 778.39	182 113.58	91 21 17

Solution:
Using A and B determine the trial coordinates P_1:

$\alpha_A=248°44′22″$ $\alpha_B=323°16′02″$
$\tan=2.570\,094$ $\tan=-0.746\,267$
$\cot=0.389\,091$ $\cot=-1.340\,002$

Using Equation (3.5a) and (3.5b):

$\cot\alpha_A-\cot\alpha_B=1.729\,093$
$E_{P1}=[(428\,905.74\times0.389\,091)-(429\,974.47\times-1.340\,002)$
$\quad\quad-182\,282.61+179\,634.85]/1.729\,093$
$\quad\quad=428\,202.68$
$N_{P1}=182\,282.61+(428\,202.68-428\,905.74)0.389\,091$
$\quad\quad=182\,009.06$

and

$N_{P1}=179\,634.85+(428\,202.68-429\,974.47)(-1.340\,02)$
$\quad\quad=182\,009.06$ (check)

Distances:

$\Delta E_{AP1}=-703.06, \quad \Delta N_{AP1}=-273.55$
Distance $AP_1=(703.06^2+273.55^2)^{1/2}=754.40$ m
$\Delta E_{BP1}=-1771.79, \quad \Delta N_{BP1}=2374.21$
Distance $BP_1=(1771.79^2+2374.21^2)^{1/2}=2962.45$ m

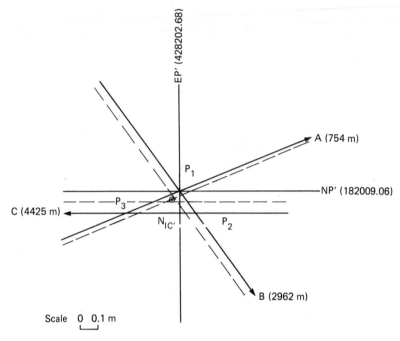

Figure 4.11 Semigraph intersection: error figure solution

A graph with P_1 as origin may now be constructed (Figure 4.11) and rays from P_1 to A and B drawn in.
Calculating the cut from C:

$E_{P1} = 428\,202.68$ $N_{P1} = 182\,009.06$
$E_C = 423\,778.39$ $N_C = 182\,113.58$
$\Delta E_{CP1} = +4424.29$ $\Delta N_{CP1} = -104.52$ Distance $= 4425.52$
$\alpha_C = 91°21'17''$, $\tan_c = -42.285$, $\cot\alpha_c = -0.023\,649$

Using Equation (4.19b):

$N_{IC'} = 182\,113.58 + [4424.29\,(-0.023\,649)]$
$\qquad = 182\,008.95$

This cutting point is now plotted and the ray through it to the point C drawn in forming the third side of the error figure $P_1P_2P_3$.
The final position of P is determined as follows, using (a) a graphical solution, or (b) a numerical solution.

(a) Graphical solution

	Distance from P	Approximate ratio of offset distances
A	754 m	1
B	2962 m	4
C	4425 m	6

Using the procedure outlined in Section 4.3 (Figure 4.10*d*), a final position is chosen to satisfy the above ratios and its coordinates scaled off:

P = 428 202.64, 182 009.02

(b) Numerical solution

$$E_{P1} = 428\,202.68 \qquad N_{P1} = 182\,009.06$$

Using B and C to coordinate P_2:

$\alpha_B = 323°16'02''$ $\qquad \alpha_C = 91°21'17''$
$\tan = -0.746\,267$ $\qquad \tan = -42.285$
$\cot = -1.340\,002$ $\qquad \cot = -0.023\,649$

Using Equations (3.5*a*) and (3.5*b*):

$\cot \alpha_B - \cot \alpha_C = -1.316\,353$
$E_{P2} = [429\,974.47\,(-1.340\,02) - 423\,778.39\,(-0.023\,649)$
$\qquad - 179\,634.85 + 182\,113.58]/ - 1.316\,353$
$\qquad = 428\,202.76$
$N_{P2} = 179\,634.85 + (428\,202.76 - 429\,974.47)(-1.340\,002)$
$\qquad = 182\,008.95$

and:

$N_{P2} = 182\,113.58 + (428\,202.76 - 423\,778.39)(-0.023\,649)$
$\qquad = 182\,008.95$ (check)

Similarly, using A and C to coordinate P_3:

$\alpha_A = 240°44'22''$ $\qquad \alpha_C = 91°21'17''$
$\tan = 2.570\,094$ $\qquad \tan = 42.285$
$\cot = 0.389\,091$ $\qquad \cot = -0.023\,649$

Using Equations (3.5*a*) and (3.5*b*):

$\cot \alpha_A - \cot \alpha_C = 0.412\,740$
$E_{P3} = [428\,905.74\,(0.389\,091) - 423\,778.39\,(-0.023\,649)$
$\qquad - 182\,282.61 + 182\,113.58]/0.412\,740$
$\qquad = 428\,202.42$
$N_{P3} = 182\,282.61 + (428\,202.42 - 428\,905.74)(0.389\,091)$
$\qquad = 182\,008.95$

and:

$$N_{P3} = 182\,113.58 + (428\,202.42 - 423\,778.39)(-0.023\,649)$$
$$= 182\,008.95 \text{ (check)}$$

We now have:

E_{P1}	428 202.68	N_{P1}	182 009.06
E_{P2}	428 202.76	N_{P2}	182 008.95
E_{P3}	428 202.42	N_{P3}	182 008.95

$\Delta E_{P1P2} = 0.08$	$\Delta N_{P1P2} = -0.11$	Bearing $P_1P_2 = 143°58'21''$
$\Delta E_{P1P3} = -0.26$	$\Delta N_{P1P3} = -0.11$	Bearing $P_1P_3 = 247°04'04''$
$\Delta E_{P2P3} = -0.34$	$\Delta N_{P2P3} = \pm 0.0$	Bearing $P_2P_3 = 270°00'00''$

Angle	*Bisection ratio*
$P_1 = 103°05'43''$	$1(A):4(B)$
$P_2 = 53\ 58\ 21$	$4(B):6(C)$
$P_3 = 22\ 55\ 56$	$1(A):6(C)$
$180\ 00\ 00$	

Angle components

P_{11}	20°37'09''	P_{12}	82°28'34''
P_{21}	21°35'20''	P_{22}	32°23'01''
P_{31}	3°16'34''	P_{32}	19°39'22''
	45 29 03		134 30 57

Bearing P_1P = Bearing $P_1P_2 + P_{12}$
$$= 143°58'21'' + 82°28'34'' = 226°26'55''$$
$$= \text{Bearing } P_1P_3 - P_{11}$$
$$= 247°04'04'' - 20°37'09'' = 226°26'55'' \text{ (check)}$$
Bearing P_2P = Bearing $P_2P_1 - P_{21}$
$$= 323°58'21'' - 21°35'20'' = 302°23'01''$$
$$= \text{Bearing } P_2P_3 + P_{22}$$
$$= 270°00'00'' + 32°23'01'' = 302°23'01'' \text{ (check)}$$
Bearing P_3P = Bearing $P_3P_2 - P_{32}$
$$= 90°00'00'' - 19°39'22'' = 70°20'38''$$
$$= \text{Bearing } P_3P_1 + P_{31}$$
$$= 67°04'04'' + 3°16'34'' = 70°20'38'' \text{ (check)}$$

To fix P from P_1 and P_2:

Bearing $P_1P = 226°26'55''$	Bearing $P_2P = 302°23'01''$
$\tan = 1.051\,889$	$\tan = -1.576\,745$
$\cot = 0.950\,671$	$\cot = -0.634\,218$

$$\cot(P_1P) - \cot(P_2P) = 1.584\,889$$

Using Equation (3.5a) and (3.5b)

$$E_P = [428\,202.68\,(0.950\,671) - 428\,202.76\,(-0.634\,218)$$
$$- 182\,009.06 + 182\,008.95]/1.584\,889$$
$$= 428\,202.64$$

$N_P = 182\,009.06 + (429\,202.64 - 428\,202.68)(0.950\,671)$
$\quad = 182\,009.02$

Also:

$N_P = 182\,008.95 + (428\,202.64 - 428\,202.76)(-0.634\,218)$
$\quad = 182\,009.02$ (check)

To fix P from P_2 and P_3:

Bearing $P_2P = 302°23'01''$ Bearing $P_3P = 70°20'38''$
$\tan = -1.576\,745$ $\tan = 2.799\,647$
$\cot = -0.634\,218$ $\cot = 0.357\,188$
$\cot(P_2P) - \cot(P_3P) = -0.991\,406$

Again using Equation (3.5a) and (3.5b):

$E_P = [428\,202.76\,(-0.634\,218) - 428\,202.42\,(0.357\,188)$
$\quad\quad - 182\,008.95 + 182\,008.95]/-0.991\,406$
$\quad = 428\,202.64$ (check)
$N_P = 182\,008.95 + (428\,202.64 - 428\,202.76)(-0.634\,218)$
$\quad = 182\,009.02$ (check)

Also:

$N_P = 182\,008.95 + (428\,202.64 - 428\,202.42)(0.357\,188)$
$\quad = 182\,009.02$ (check)

Program 4.17 Semigraphic intersection

```
           EASTINGS              NORTHINGS              BEARING
DATUM1    428905.74             182282.61           248.739444444
DATUM2    429974.47             179634.85           323.267222222
DATUM3    423778.39             182113.58            91.3547222222

PROVISIONAL VALUES OF P
    EASTINGS 428202.678282NORTHINGS 182009.055128

CHECK  182009.055128

NORTHINGS OF CUT=182008.951022
INTERCEPT=0.104105759412


100 PRINT "PROGRAM 4.17"
110 REM THIS PROGRAM CALCULATES THE CUTS REQUIRED FOR A SEMI-GRAPHIC
120 REM INTERSECTION BASED ON Q DATUM POINTS
130 INIT
140 SET DEGREES
150 DIM E(10),N(10),B(10)
160 PRINT "ENTER NUMBER OF DATUM POINTS";
170 INPUT Q
180 FOR I=1 TO Q
190 B(I)=0
200 PRINT "ENTER DEGREES, MINS, SECS OF BEARING B";I;"=";
210 INPUT D,M,S
220 B(I)=D+M/60+S/3600
230 PRINT "ENTER EASTINGS,NORTHINGS OF DATUM POINT";I;"=";
240 INPUT E(I),N(I)
250 NEXT I
260 PRINT @4:"         ";"EASTINGS ";"                  ";"NORTHINGS";
```

```
270 PRINT @4:"                    ";"BEARING"
280 FOR I=1 TO Q
290 PRINT @4:"DATUM";I;"   ";E(I);"                ";N(I);"            ";
300 PRINT @4:B(I)
310 NEXT I
320 PRINT @4:
330 E(10)=0
340 E(10)=N(1)-N(2)-E(1)*(1/TAN(B(1)))+E(2)*(1/TAN(B(2)))
350 E(10)=E(10)/(1/TAN(B(2))-1/TAN(B(1)))
360 N(10)=N(1)+(E(10)-E(1))*(1/TAN(B(1)))
370 PRINT @4:"PROVISIONAL VALUES OF P "
380 PRINT @4:"        ";"EASTINGS ";E(10);"NORTHINGS ";N(10)
390 Z=N(2)+(E(10)-E(2))*(1/TAN(B(2)))
400 PRINT @4:
410 PRINT @4:"CHECK";"   ";Z
420 PRINT @4:
430 FOR I=3 TO Q
440 E(9)=0
450 N(9)=0
460 E(9)=E(10)-E(I)
470 N(9)=N(10)-N(I)
480 IF E(9)>0 THEN 520
490 IF N(9)>0 THEN 550
500 IF ABS(E(9))>ABS(N(9)) THEN 590
510 GO TO 610
520 IF N(9)>0 THEN 550
530 IF ABS(E(9))>ABS(N(9)) THEN 590
540 GO TO 610
550 IF ABS(E(9))>ABS(N(9)) THEN 570
560 GO TO 630
570 N(8)=E(9)*(1/TAN(B(I)))
580 GO TO 650
590 N(8)=E(9)*(1/TAN(B(I)))
600 GO TO 650
610 E(8)=N(9)*TAN(B(I))
620 GO TO 710
630 E(8)=N(9)*TAN(B(I))
640 GO TO 710
650 N(7)=N(I)+N(8)
660 PRINT @4:"NORTHINGS OF CUT=";N(7)
670 N(6)=N(10)-N(7)
680 PRINT @4:"INTERCEPT=";N(6)
690 NEXT I
700 GO TO 760
710 E(7)=E(I)+E(8)
720 PRINT @4:"E(7)=";E(7)
730 E(6)=E(10)-E(7)
740 PRINT @4:"E(6)=";E(6)
750 NEXT I
760 STOP
770 END
```

4.4 Resection

The position of a point may be determined by observing at that point to at least three previously fixed datum stations (providing that the point is not concyclic with the datum stations). The only required observations are the measurement of the angles subtended at the resected point by the three control points as illustrated in Figure 4.12.

The main problem stems from the fact that the bearings of the three lines observed are initially unknown. Several alternative approaches to the solution of this problem may be used. The following formula however provides the direct determination of the bearing of the

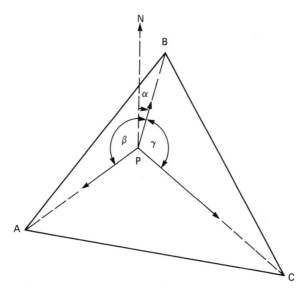

Figure 4.12 Resection: point inside the datum triangle

central line of the three observed stations and lends itself to a computerized solution.

$$\tan \alpha_{PB} = \frac{(E_B - E_A)\cot \beta + (E_B - E_C)\cot \gamma - N_A + N_C}{(N_B - N_A)\cot \beta + (N_B - N_C)\cot \gamma + E_A - E_C}$$ (4.20)

$$\alpha_{BP} = \alpha_{PB} \pm 180°$$

$$\alpha_{PC} = \alpha_{PB} + \gamma \quad \therefore \quad \alpha_{CP} = (\alpha_{PB} + \gamma) \pm 180°$$

and:

$$\alpha_{PA} = \alpha_{PB} - \beta \quad \therefore \quad \alpha_{AP} = (\alpha_{PB} - \beta) \pm 180°$$

The signs of the numerator and denominator in the above expression indicate in which quadrant α_{PB} lies.

Using the bearing intersection formula (Equations (3.5) or (3.6)) and any pair of datum points a unique position of P may be determined. The coordinate values thus derived may be verified by using one or both of the remaining datum pairs.

Example 4.17
Referring to Figure 4.13 the observed angles at P are $\beta = 45°10'58''$ and $\gamma = 87°14'10''$.

86   Control surveys

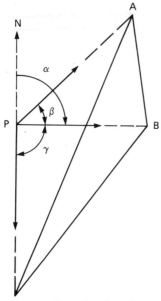

C   *Figure 4.13* Resection: point outside the datum triangle

Coordinates:

	E	(m)	N
A	356 442.71		148 778.96
B	356 788.67		144 328.27
C	351 240.22		138 628.80

Determine the coordinates of P.

Solution:
Using Equation (4.20),

$$\tan \alpha_{PB} = \frac{M}{N}$$

$$M = (356\,788.67 - 356\,442.71)(0.993\,640)$$
$$+ (356\,788.67 - 351\,240.22)(0.048\,276)$$
$$- 148\,778.96 + 138\,628.80$$
$$= -9538.543$$
$$N = (144\,328.27 - 148\,778.96)(0.993\,640)$$
$$+ (144\,328.27 - 138\,628.80)(0.048\,276)$$
$$+ 356\,442.71 - 351\,240.22$$
$$= 1055.254$$

$$\tan \alpha_{PB} = \frac{-9538.543}{1055.254} = -9.039\,096\,7$$
$$\alpha_{PB} = 96°18'47''$$

Hence:

$$\alpha_{PA} = 96°18'47'' - 45°10'58'' = 51°07'49''$$
$$\alpha_{PC} = 96°18'47'' + 87°14'10'' = 183°32'57''$$

	tan	cot
$\alpha_{AP} = 231°07'49''$	1.240 655	0.806 026
$\alpha_{BP} = 276°18'47''$	-9.039 003	-0.110 632
$\alpha_{CP} = 03°32'57''$	0.062 024	16.122 792

Using A, B and Equations (3.5a) and (3.5b):

$$\cot \alpha_{AP} - \cot \alpha_{BP} = 0.916\,658$$
$$E_P = [356\,442.71\,(0.806\,026) - 356\,788.67\,(-0.110\,632)$$
$$- 148\,778.96 + 144\,328.27]/0.916\,658$$
$$= 351\,629.12$$
$$N_P = 148\,778.96 + (351\,629.12 - 356\,442.71)(0.806\,026)$$
$$= 144\,899.08$$

and:

$$N_P = 144\,328.27 + (351\,629.12 - 356\,788.67)(-0.110\,632)$$
$$= 144\,899.08 \text{ (check)}$$

Using B, C and Equation (3.6a) and (3.6b):

$$\tan \alpha_{BP} - \tan \alpha_{CP} = -9.101\,027$$
$$N_P = [144\,328.27\,(-9.039\,003) - 138\,628.80\,(0.062\,024)$$
$$- 356\,788.67 + 351\,240.22]/-9.101\,027$$
$$= 144\,899.08 \text{ (check)}$$
$$E_P = 356\,788.67 + (144\,899.08 - 144\,328.27)(-9.039\,003)$$
$$= 351\,629.12 \text{ (check)}$$

and:

$$E_P = 351\,240.22 + (144\,899.08 - 138\,628.80)(0.062\,024)$$
$$= 351\,629.12 \text{ (check)}$$

Program 4.18 Resection

```
EASTINGS OF  A=  356442.71   NORTHINGS OF A=  148778.96
             B=  356788.67                 B=  144328.27
             C=  351240.22                 C=  138628.8

ANGLE BETA=  45.1827777778  ANGLE GAMMA=87.2361111111

BEARING P TO B=96.3130026095
BEARING P TO A=51.1302248317
```

```
BEARING P TO C=183.549113721
EASTINGS OF P=  351629.121672   NORTHINGS OF P=   144899.074639
CHECK=144899.074639

CHECKS

EASTINGS OF P=  351629.121672   NORTHINGS OF P=   144899.074639
CHECK=144899.074639


100 PRINT "PROGRAM 4.18"
110 REM THIS PROGRAM SOLVES THE RESECTION PROBLEM AND DETERMINES THE
120 REM COORDINATES OF A RESECTED POINT GIVEN THE COORDINATES OF THREE
130 REM OBSERVED DATUM POINTS AND THE TWO OBSERVED ANGLES AT THE
140 REM RESECTED POINT
150 INIT
160 SET DEGREES
170 PRINT "ENTER EASTINGS, NORTHINGS OF A=";
180 INPUT E1,N1
190 PRINT "ENTER EASTINGS, NORTHINGS OF B=";
200 INPUT E2,N2
210 PRINT "ENTER EASTINGS, NORTHINGS OF C=";
220 INPUT E3,N3
230 PRINT "ENTER DEGREES,MINS,SECS OF ANGLE BETA=";
240 INPUT D1,M1,S1
250 PRINT "ENTER DEGREES,MINS,SECS OF ANGLE GAMMA=";
260 INPUT D2,M2,S2
270 A1=D1+M1/60+S1/3600
280 A2=D2+M2/60+S2/3600
290 M=(E2-E1)*(1/TAN(A1))+(E2-E3)*(1/TAN(A2))-N1+N3
300 N=(N2-N1)*(1/TAN(A1))+(N2-N3)*(1/TAN(A2))+E1-E3
310 IF N<0 THEN 370
320 IF M<0 THEN 350
330 B2=ATN(M/N)
340 GO TO 410
350 B2=180-ABS(ATN(M/N))
360 GO TO 410
370 IF M<0 THEN 400
380 B2=360-ABS(ATN(M/N))
390 GO TO 410
400 B2=ATN(M/N)+180
410 B3=B2+A2
420 B1=B2-A1
430 E4=N1-N2-E1*(1/TAN(B1))+E2*(1/TAN(B2))
440 E4=E4/(1/TAN(B2)-1/TAN(B1))
450 N4=N1+(E4-E1)*(1/TAN(B1))
460 Z1=N2+(E4-E2)*(1/TAN(B2))
470 PRINT @4:"EASTINGS OF A=";"   ";E1;"   ";"NORTHINGS OF A=";"   ";N1
480 PRINT @4:"            B=";"   ";E2;"   ";"            B=";"   ";N2
490 PRINT @4:"            C=";"   ";E3;"   ";"            C=";"   ";N3
500 PRINT @4:
510 PRINT @4:"ANGLE BETA=";"   ";A1;"   ";"ANGLE GAMMA=";A2
520 PRINT @4:
530 PRINT @4:"BEARING P TO B=";B2
540 PRINT @4:"BEARING P TO A=";B1
550 PRINT @4:"BEARING P TO C=";B3
560 PRINT @4:"EASTINGS OF P=";"   ";E4;"   ";"NORTHINGS OF P=";"   ";N4
570 PRINT @4:"CHECK=";Z1
580 E5=N2-N3-E2*(1/TAN(B2))+E3*(1/TAN(B3))
590 E5=E5/(1/TAN(B3)-1/TAN(B2))
600 N5=N2+(E5-E2)*(1/TAN(B2))
610 Z2=N3+(E5-E3)*(1/TAN(B3))
620 PRINT @4:
630 PRINT @4:"CHECKS"
640 PRINT @4:
650 PRINT @4:"EASTINGS OF P=";"   ";E5;"   ";"NORTHINGS OF P=";"   ";N5
660 PRINT @4:"CHECK=";Z2
670 STOP
680 END
```

4.5 Traversing

Traversing is one of the most commonly used methods for providing horizontal survey control. Figures 4.14(a) and (b) illustrate two common traverse configurations. Traverse (a) is typical of the most general case where the traverse commences at a datum point A with opening bearing orientation to a second datum point B. The lengths of the traverse lines and the angles at each traverse point (measured clockwise from the back station to the forward station) are determined and the traverse closes at datum point C with closing bearing orientation to datum point D.

In traverse (b) only two datum points are used, AC providing the opening bearing and the reciprocal CA, the closing bearing.

Figure 4.14a Control traverse (a)

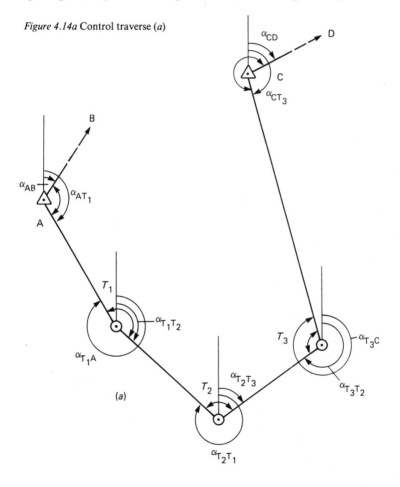

(a)

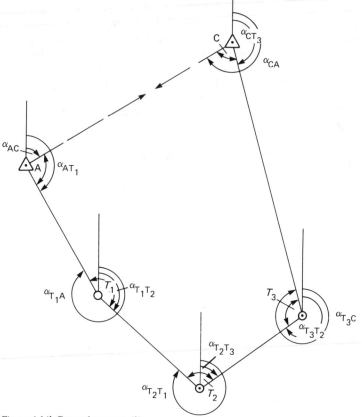

Figure 4.14b Control traverse (*b*)

4.5.1 Reduction of distances

Before commencing the computation of a traverse the measured distances must first be reduced by the application of relevant corrections noted in Section 4.1.1.

4.5.2 Reduction of bearings

The first step in the computation of a traverse is to apply a check to the angular observations to ensure their consistency, to determine whether discrepancies are within specified tolerances, and to adjust such discrepancies between the angles or bearings.

In a traverse forming a totally closed circuit (Figure 4.14(*b*)) the internal angles, *n* in number, should sum $(2n-4)$ right angles. Alternatively, the external angles should sum $(2n+4)$ right angles.

Where a traverse uses different lines for the initial and closing bearings (Figure 4.14(a)) the sum of the observed clockwise angles measured at all stations, including the terminal datum station, minus the difference between the closing and initial bearings (increased by 360° if the former is smaller) should equal $(n-1) \times 180°$ where n is the number of stations including the two terminal datum stations.

Providing the angular misclosure is within the specified tolerance it may then be distributed uniformly between each angle following which bearings may be derived using the initial datum bearing and the adjusted angles.

Alternatively, preliminary directions for each traverse leg may be computed directly using the initial datum bearing and the observed traverse angles. The difference between the bearing thus derived for the final orientation ray and its datum value indicates the bearing misclosure. Final bearings may then be adjusted by applying corrections cumulatively to the provisional directions. The following examples illustrate the bearing adjustment procedures.

Example 4.18
The mean observed internal angles of the traverse shown in Figure 4.14(b) are as follows:

A	89°47′01″
T1	162 00 47
T2	104 23 01
T3	107 59 11
C	75 49 45

The bearing of the line AC is 60°00′00″.
Derive adjusted bearings for the traverse lines.

Solution:

	Observed angle	Adjustment	Adjusted angle	Adjusted bearing			
	° ′ ″		° ′ ″	AC	60°	00′	00″
A	89 47 01	+03″	89 47 04				
				AT$_1$	149	47	04
T$_1$	162 00 47	+03	162 00 50	((T$_1$A)	329	47	04)
				T$_1$T$_2$	131	47	54
T$_2$	104 23 01	+03	104 23 04	((T$_2$T$_1$)	311	47	54)
				T$_2$T$_3$	56	10	58
T$_3$	107 59 11	+03	107 59 14	((T$_3$T$_2$)	236	10	58)
				T$_3$C	344	10	12
C	75 49 45	+03	75 49 48	((CT$_3$)	164	10	12)
Sum	539 59 45			CA	240	00	00

The sum should be $(2n-4)\,90°$ where $n=5$, i.e. 540°. Hence, misclosure $-15″$, adjustment $=15/5=+3″$ per station.

Program 4.19 Traverse bearing adjustment (1)

```
ANGLE1=89.7836111111
ANGLE2=162.013055556
ANGLE3=104.383611111
ANGLE4=107.986388889
ANGLE5=75.8291666667

SUM OF INTERNAL ANGLES=539.995833333
SUM SHOULD EQUAL        540
MISCLOSURE             =-0.0041666666666933
ADJUSTMENT             =8.333333339E-4

ADJUSTED ANGLES

ANGLE1=89.7844444444

ANGLE2=162.013888889

ANGLE3=104.384444444

ANGLE4=107.987222222

ANGLE5=75.83
SUM OF ADJUSTED ANGLES=540

ADJUSTED BEARINGS

INITIAL DATUM BEARING=60 0 0
BEARING OF LEG1=149 47 4
BEARING OF LEG2=131 47 54
BEARING OF LEG3=56 10 58
BEARING OF LEG4=344 10 12
CLOSING BEARING=239 59 60

SHOULD BE        240 0 0
```

```
100 PRINT "PROGRAM 4.19"
110 REM THIS PROGRAM ADJUSTS THE OBSERVED ANGLES OF A CLOSED TRAVERSE
120 REM AND DERIVES REDUCED BEARINGS FOR THE INDIVIDUAL LEGS OF THE
130 REM TRAVERSE
140 INIT
150 SET DEGREES
160 PRINT "ENTER DEGREES,MINS,SECS OF INITIAL BEARING=";
170 INPUT E1,F1,G1
180 PRINT "ENTER DEGREES,MINS,SECS OF FINAL BEARING  =";
190 INPUT E2,F2,G2
200 PRINT "ENTER NUMBER OF STATIONS IN TRAVERSE FROM WHICH OBSERVATIONS"
210 PRINT "HAVE BEEN MADE=";
220 INPUT N
230 DIM D(N),M(N),S(N),A(N),B(N)
240 Z1=0
250 Z2=0
260 FOR I=1 TO N
270 PRINT "ENTER DEGREES,MINS,SECS OF INTERNAL OBSERVED ANGLE";I
280 INPUT D(I),M(I),S(I)
290 A(I)=D(I)+M(I)/60+S(I)/3600
300 PRINT @4:"ANGLE";I;"=";A(I)
310 Z1=Z1+A(I)
320 NEXT I
330 PRINT @4:
340 PRINT @4:"SUM OF INTERNAL ANGLES=";Z1
350 Q=(2*N-4)*90
360 PRINT @4:"SUM SHOULD EQUAL       ";Q
370 V=Z1-Q
380 PRINT @4:"MISCLOSURE            =";V
390 W=-(V/N)
400 PRINT @4:"ADJUSTMENT            =";W
410 PRINT @4:
420 PRINT @4:"ADJUSTED ANGLES"
```

```
430 FOR I=1 TO N
440 A(I)=A(I)+W
450 PRINT @4:
460 PRINT @4:"ANGLE";I;"=";A(I)
470 Z2=Z2+A(I)
480 NEXT I
490 PRINT @4:"SUM OF ADJUSTED ANGLES=";Z2
500 PRINT @4:
510 PRINT @4:"ADJUSTED BEARINGS"
520 X1=E1+F1/60+G1/3600
530 X2=E2+F2/60+G2/3600
540 B(1)=X1+A(1)
550 IF B(1)<360 THEN 570
560 B(1)=B(1)-360
570 O=B(1)
580 GOSUB 790
590 PRINT @4:
600 PRINT @4:"INITIAL DATUM BEARING=";E1;F1;G1
610 PRINT @4:"BEARING OF LEG1=";P;INT(Q);INT(R)
620 FOR I=2 TO N-1
630 B(I)=B(I-1)+180+A(I)
640 IF B(I)<360 THEN 660
650 B(I)=B(I)-360
660 O=B(I)
670 GOSUB 790
680 PRINT @4:"BEARING OF LEG";I;"=";P;INT(Q);INT(R)
690 NEXT I
700 B(N)=B(N-1)+180+A(N)
710 IF B(N)<360 THEN 730
720 B(N)=B(N)-360
730 O=B(N)
740 GOSUB 790
750 PRINT @4:"CLOSING BEARING=";P;INT(Q);INT(R)
760 PRINT @4:
770 PRINT @4:"SHOULD BE        ";E2;F2;G2
780 GO TO 830
790 P=INT(O)
800 Q=(O-INT(O))*60
810 R=(Q-INT(Q))*60+0.5
820 RETURN
830 STOP
840 END
```

Example 4.19
Use the same data as in Example 4.18.

Solution:

	Observed angle		Unadjusted bearing	Adjust-ment	Adjusted bearing			
	° ′ ″		° ′ ″		AC	60°	00′	00″
A	89 47 01	AT_1	149 47 01	+3		149	47	04
T1	162 00 47	$(T_1A$	329 47 01)					
		T_1T_2	131 47 48	+6		131	47	54
T2	104 23 01	$(T_2T_1$	311 47 48)					
		T_2T_3	56 10 49	+9		56	10	58
T3	107 59 11	$(T_3T_2$	236 10 49)					
		T_3C	344 10 00	+12		344	10	12
C	75 49 45	$(CT_3$	164 10 00)					
		CA	239 59 45	+15	CA	240	00	·00

	should be	=	240 00 00
∴	misclosure	=	−15″, adjustment = 15/5 = +3″ per station.

Program 4.20 Traverse bearing adjustment (2)

```
ANGLE1=89.7836111111
ANGLE2=162.013055556
ANGLE3=104.383611111
ANGLE4=107.986388889
ANGLE5=75.8291666667


INITIAL DATUM BEARING=60 0 0
UNADJUSTED FORWARD BEARINGS


LEG 1=149.783611111
LEG2=131.796666667
LEG3=56.1802777778
LEG4=344.166666667
UNADJUSTED FINAL BEARING=239.995833333
SHOULD BE 240
MISCLOSURE=-0.00416666666933
ADJUSTMENT=8.333333339E-4


ADJUSTED FORWARD BEARINGS


LEG1=149 47 4


LEG2=131 47 54


LEG3=56 10 58


LEG4=344 10 12


LEG5=239 59 60




100 PRINT "PROGRAM 4.20"
110 REM THIS PROGRAM USES THE OBSERVED ANGLES OF A TRAVERSE TO DETERMINE
120 REM THE UNADJUSTED FORWARD BEARINGS WHICH ARE USED TO DERIVE THE
130 REM ADJUSTED FINAL BEARINGS
140 INIT
150 SET DEGREES
160 PRINT "ENTER DEGREES,MINS,SECS OF INITIAL DATUM BEARING=";
170 INPUT E1,F1,G1
180 PRINT "ENTER DEGREES,MINS,SECS OF FINAL DATUM BEARING=";
190 INPUT E2,F2,G2
200 PRINT "ENTER NUMBER OF STATIONS IN TRAVERSE FROM WHICH OBSERVATIONS"
210 PRINT "HAVE BEEN MADE=";
220 INPUT N
230 DIM D(N),M(N),S(N),A(N),B(N)
240 FOR I=1 TO N
250 PRINT "ENTER DEGREES,MINS,SECS OF INTERNAL OBSERVED ANGLE";I
260 INPUT D(I),M(I),S(I)
270 A(I)=D(I)+M(I)/60+S(I)/3600
280 PRINT @4:"ANGLE";I;"=";A(I)
290 NEXT I
300 X1=E1+F1/60+G1/3600
310 X2=E2+F2/60+G2/3600
320 PRINT @4:
330 PRINT @4:"INITIAL DATUM BEARING=";E1;F1;G1
340 B(1)=X1+A(1)
350 IF B(1)<360 THEN 370
360 B(1)=B(1)-360
370 PRINT @4:"UNADJUSTED FORWARD BEARINGS"
380 PRINT @4:
390 PRINT @4:"LEG 1=";B(1)
400 FOR I=2 TO N-1
```

```
410 B(I)=B(I-1)+180+A(I)
420 IF B(I)<360 THEN 440
430 B(I)=B(I)-360
440 PRINT @4:"LEG";I;"=";B(I)
450 NEXT I
460 B(N)=B(N-1)+180+A(N)
470 IF B(N)<360 THEN 490
480 B(N)=B(N)-360
490 PRINT @4:"UNADJUSTED FINAL BEARING=";B(N)
500 PRINT @4:"SHOULD BE ";X2
510 V=B(N)-X2
520 PRINT @4:"MISCLOSURE=";V
530 W=-(V/N)
540 PRINT @4:"ADJUSTMENT=";W
550 PRINT @4:
560 PRINT @4:"ADJUSTED FORWARD BEARINGS"
570 FOR I=1 TO N
580 B(I)=B(I)+W*I
590 O=B(I)
600 GOSUB 650
610 PRINT @4:
620 PRINT @4:"LEG";I;"=";P;INT(Q);INT(R)
630 NEXT I
640 GO TO 690
650 P=INT(O)
660 Q=(O-INT(O))*60
670 R=(Q-INT(Q))*60+0.5
680 RETURN
690 STOP
700 END
```

Example 4.20
The mean observed clockwise angles of the traverse shown in Figure
4.14(*a*) are as follows:

A	119°17′13″
T1	162 00 47
T2	104 23 01
T3	107 59 11
C	256 51 06

Datum bearings are:

AB	30 29 43
CD	61 01 26

Derive adjusted bearings for the traverse lines.

Solution:

	Angle	Adjustment	Adjusted bearing
B	° ′ ″		° ′ ″
			210 29 43
A	119 17 13	+5	
			149 47 01
T$_1$	162 00 47	+5	
			131 47 53
T$_2$	104 23 01	+5	
			56 10 59
T$_3$	107 59 11	+5	
			344 10 15
C	256 51 06	+5	
			61 01 26
D			
Sum	750 31 18		

Sum should be $[(n-1)180°] + (61°01'26'' - 30°29'43'')$ where $n = 5$, i.e. $750°31'43''$. $\therefore$ misclosure $= -25''$, adjustment $= +25/5 = +5''$ per station.

Program 4.21 Traverse bearing adjustment (3)

```
ANGLE1=119.286944444
ANGLE2=162.013055556
ANGLE3=104.383611111
ANGLE4=107.986388889
ANGLE5=256.851666667

INITIAL DATUM BEARING=30.4952777778
FINAL DATUM BEARING  =61.0238888889
DIFFERENCE INITIAL AND FINAL DATUM BEARINGS=30.5286111111

SUM OBSERVED ANGLES  =750.521666667
SHOULD BE             750.528611111
MISCLOSURE=-0.00694444444889
ADJUSTMENT=0.00138888888978

ADJUSTED FORWARD BEARINGS
LEG1=149 47 1
LEG2=131 47 53
LEG3=56 10 59
LEG4=344 10 15
DEDUCED FINAL BEARING=61 1 26
SHOULD BE (CHECK)      61 1 26


100 PRINT "PROGRAM 4.21"
110 REM THIS PROGRAM DETERMINES THE BEARINGS OF A TRAVERSE WHICH DOES
120 REM NOT FORM A CLOSED POLYGON (REF EXAMPLE 4.20 AND FIG 4.14(A)
130 INIT
140 SET DEGREES
150 PRINT "ENTER DEGREES,MINS,SECS OF INITIAL DATUM BEARING=";
160 INPUT E1,F1,G1
170 PRINT "ENTER DEGREES,MINS,SECS OF FINAL DATUM BEARING=";
180 INPUT E2,F2,G2
190 PRINT "ENTER NUMBER OF STATIONS IN TRAVERSE FROM WHICH OBSERVATIONS"
200 PRINT "HAVE BEEN MADE=";
210 INPUT N
220 DIM D(N),M(N),S(N),A(N),B(N)
```

```
230 Z1=0
240 FOR I=1 TO N
250 PRINT "ENTER DEGREES,MINS,SECS OF CLOCKWISE OBSERVED ANGLE";I
260 INPUT D(I),M(I),S(I)
270 A(I)=D(I)+M(I)/60+S(I)/3600
280 PRINT @4:"ANGLE";I;"=";A(I)
290 Z1=Z1+A(I)
300 NEXT I
310 X1=E1+F1/60+G1/3600
320 X2=E2+F2/60+G2/3600
330 PRINT @4:
340 PRINT @4:"INITIAL DATUM BEARING=";X1
350 PRINT @4:"FINAL DATUM BEARING  =";X2
360 X3=X1-X2
370 IF X3>0 THEN 390
380 X3=-X3
390 PRINT @4:"DIFFERENCE INITIAL AND FINAL DATUM BEARINGS=";X3
400 Z2=X3+(N-1)*180
410 PRINT @4:
420 PRINT @4:"SUM OBSERVED ANGLES  =";Z1
430 PRINT @4:"SHOULD BE            ";Z2
440 V=Z1-Z2
450 W=-(V/N)
460 PRINT @4:"MISCLOSURE=";V
470 PRINT @4:"ADJUSTMENT=";W
480 B(1)=X1+A(1)+W
490 IF B(1)<360 THEN 530
500 B(1)=B(1)-360
510 IF B(1)<360 THEN 530
520 B(1)=B(1)-360
530 O=B(1)
540 PRINT @4:
550 PRINT @4:"ADJUSTED FORWARD BEARINGS"
560 GOSUB 780
570 PRINT @4:"LEG1=";P;INT(Q);INT(R)
580 FOR I=2 TO N-1
590 B(I)=B(I-1)+180+A(I)+W
600 IF B(I)<360 THEN 640
610 B(I)=B(I)-360
620 IF B(I)<360 THEN 640
630 B(I)=B(I)-360
640 O=B(I)
650 GOSUB 780
660 PRINT @4:"LEG";I;"=";P;INT(Q);INT(R)
670 NEXT I
680 B(N)=B(N-1)+180+A(N)+W
690 IF B(N)<360 THEN 730
700 B(N)=B(N)-360
710 IF B(N)<360 THEN 730
720 B(N)=B(N)-360
730 O=B(N)
740 GOSUB 780
750 PRINT @4:"DEDUCED FINAL BEARING=";P;INT(Q);INT(R)
760 PRINT @4:"SHOULD BE (CHECK)    ";E2;F2;G2
770 GO TO 820
780 P=INT(O)
790 Q=(O-INT(O))*60
800 R=(Q-INT(Q))*60+0.5
810 RETURN
820 STOP
830 END
```

4.5.3 Traverse computation

1. The adjusted bearings (α) and reduced distances (l) may now be used to determine the partial coordinates (ΔE and ΔN) for each leg of the traverse in turn, that is:

$$\Delta E_{ij} = l_{ij} \sin \alpha_{ij} \qquad (4.21a)$$

$$\Delta N_{ij} = l_{ij} \cos \alpha_{ij} \tag{4.21b}$$

where i and j represent the two terminal stations of any traverse leg.

2. In a closed traverse the sum of the partial eastings coordinates and the sum of the partial northings coordinates should equal the respective differences in eastings and northings between the two terminal datum points. The amount by which the traverse fails to satisfy this condition is a measure of the accuracy of the traverse and may be demonstrated by the ratio:

$$(dE^2 + dN^2)^{1/2} / L$$

where L is the total length of all traverse legs and dE and dN are the differences between the sums of the partial coordinates of all traverse legs ($\Sigma \Delta E_{ij}, \Sigma \Delta N_{ij}$) and the respective differences in eastings and northings of the two terminal datum points. Figure 4.15 illustrates the misclosure diagrammatically, Y being the terminal datum point and Y' its position as derived from the

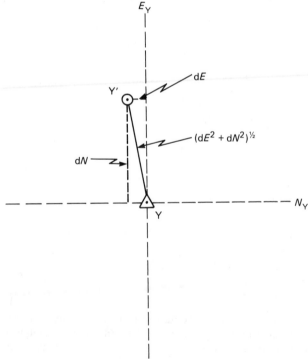

Figure 4.15 Traverse misclosure

summation of the initial computed partial coordinates for each leg of the traverse.

The corrections to be applied to the partial coordinates to make their sums agree with the datum coordinate differences may be determined in several alternative ways.

(a) One commonly used method (the Bowditch method) is that in which the corrections (e and n) to the partial coordinates of an individual leg (ΔE_{ij} and ΔN_{ij}) bears the same proportion to the total correction (dE or dN) that the length of the individual leg (l_{ij}) bears to the total length of the traverse (L), i.e. for any leg of the traverse:

$$e_{ij} = \frac{l_{ij}}{L} dE \qquad n_{ij} = \frac{l_{ij}}{L} dN \qquad (4.22)$$

Hence the *adjusted* partial coordinates to be used for the final coordination of the new traverse points are:

$$\Delta E'_{ij} = \Delta E_{ij} \pm e_{ij} \qquad \Delta N'_{ij} = \Delta N_{ij} \pm n_{ij} \qquad (4.23)$$

(b) An alternative method (the McCaw method) provides corrections (e and n) to the initial partial coordinates of an individual traverse leg which are in the same proportion to the total corrections (dE and dN) as the partial coordinates themselves (ΔE and ΔN) are to the sum of all partial coordinates ($\Sigma \Delta E$ and $\Sigma \Delta N$) *irrespective of their sign* (i.e. the sum of the absolute values of the partial coordinates).

Thus for any leg of the traverse:

$$e_{ij} = \frac{\Delta E_{ij}}{\Sigma \Delta E} dE \qquad n_{ij} = \frac{\Delta N_{ij}}{\Sigma \Delta N} dN \qquad (4.24)$$

$$\Delta E'_{ij} = \Delta E_{ij} \pm e_{ij} \qquad \Delta N'_{ij} = \Delta N_{ij} \pm n_{ij} \qquad (4.25)$$

3. Having applied the corrections to the initial partial coordinates, the final coordinates of each new traverse point are obtained by the successive application of the corrected partial coordinates to the coordinates of the previous point in the traverse.

A check that the corrections have been correctly calculated and applied is provided by verifying that the coordinates of the final terminal datum point (as computed) are equal to its datum values.

Example 4.21
Tables 4.5 and 4.6 illustrate typical traverse computations using the reduced bearings from Example 4.20 and applying in turn each of the above methods of adjustment.

Table 4.5 Method 1 See page 102 for adjustment computation

1 Station	2 Bearing (α) ° ′ ″	3 Distance (l)	4 ΔE	5 e	6 ΔN	7 n	8 Coordinates Eastings	8 Coordinates Northings
B	210 29 43						1284.531	2067.219
A	149 47 01	366.127	184.260	+0.006	−316.382	−0.032	184.266	−316.414
T_1	131 47 53	344.574	256.879	+0.006	−229.661	−0.030	1468.797	1750.805
T_2	56 10 59	317.641	263.902	+0.006	176.780	−0.028	256.885	−229.691
T_3	344 10 15	693.558	−189.182	+0.012	667.258	−0.060	1725.682	1521.114
C							263.908	176.752
D	61 01 26						1989.590	1697.866
							−189.170	667.198
							1800.420	2365.064

$L = 1721.900$ $\Sigma\Delta E = 515.859$ $dE = -0.030$ $\Sigma\Delta N = 297.995$ $dN = 0.150$ $\Delta E_{AC} = 515.889$ $\Delta N_{AC} = 297.845$

Table 4.6 Method 2 See page 102 for adjustment computation

1 Station	2 Bearing (α) ° ′ ″	3 Distance (l)	4 ΔE	5 e	6 ΔN	7 n	8 Coordinates Eastings	Northings
B	210 29 43						1284.531	2067.219
		366.127	184.260	+0.006	−316.382	−0.034	184.266	−316.416
A	149 47 01						1468.797	1750.803
		344.574	256.879	+0.009	−229.661	−0.025	256.888	−229.686
T₁	131 47 53						1725.685	1521.117
		317.641	263.902	+0.009	176.780	−0.019	263.911	176.761
T₂	56 10 59						1989.596	1697.878
		693.558	−189.182	+0.006	667.258	−0.072	−189.176	667.186
T₃	344 10 15						1800.420	2365.064
C	61 01 26							
D								
			ΣΔE = 894.223	dE = −0.030	ΣΔN = 1390.081	dN = 0.150	ΔE_AC = 515.889	ΔN_AC = 297.845

The traverse misclosure $= (0.03^2 + 0.15^2)^{1/2} = 0.153$. As a proportion of the total length $= 0.153/1721.9 = 1/11257$.

The corrections (e and n) to the partial coordinates are derived as follows:

Method 1 (Table 4.5)

For eastings, $e = \dfrac{dE}{L}l_{ij}$ and for leg AT_1, $e = \dfrac{0.030}{1721.9} \times 366.127$

$= 0.006$

For northings, $n = \dfrac{dN}{L}l_{ij}$ and for leg AT_1, $n = \dfrac{0.150}{1721.9} \times 366.127$

$= 0.031$

Method 2 (Table 4.5)

For eastings, $e = \dfrac{dE}{\Sigma\Delta E}\Delta E_{ij}$ where $\Sigma\Delta E$ is the sum of the absolute values of the partial easting coordinates.

For leg AT_1, $e = \dfrac{0.030}{894.223} \times 184.260 = 0.006$

For northings, $n = \dfrac{dN}{\Sigma\Delta N}\Delta N_{ij}$ and for leg AT_1,

$n = \dfrac{0.150}{1390.081} \times 316.382$

$= 0.034$

Programs 4.22 and 4.23 Traverse adjustment

BEARING	DISTANCE	DIFF EASTINGS	DIFF NORTHINGS
149.783611111	366.127	184.259689366	-316.381647704
131.798055556	344.574	256.87944126	-229.661041829
56.1830555556	317.641	263.902468972	176.780348884
344.170833333	693.558	-189.181838555	667.257769775

SUM PARTIAL EASTINGS=515.859761043 SUM PARTIAL NORTHINGS=297.995429126
SHOULD BE   515.889   297.845

MISCLOSURE EASTINGS=-0.0292389574606   MISCLOSURE NORTHINGS=0.150429126235

	ADJUSTMENT		CORRECTED PARTIAL COORDINATES	
	EASTINGS	NORTHINGS	EASTINGS	NORTHINGS

MCCAW ADJUSTMENT

	ADJUSTMENT EASTINGS	ADJUSTMENT NORTHINGS	CORRECTED EASTINGS	CORRECTED NORTHINGS
A TO 1	0.00602484903573	-0.0342375885923	184.265714215	-316.415885292
1 TO 2	0.00839934040537	-0.0248530226797	256.8878406	-229.685894852
2 TO 3	0.00862897653406	-0.0191304802293	263.911097948	176.761218404
3 TO 4	0.00618579148546	-0.072208034734	-189.175652763	667.18556174

FINAL COORDINATES

FIRST TERMINAL	1284.531	2067.219
1	1468.79671422	1750.80311471

```
   2                1725.68455482     1521.11721986
   3                1989.59565276     1697.87843826
FINAL TERMINAL      1800.42           2365.064


100 PRINT "PROGRAM 4.22 AND 4.23"
110 REM THIS PROGRAM DETERMINES THE ADJUSTED COORDINATES OF POINTS
120 REM INCLUDED IN A CLOSED TRAVERSE. THE PROGRAM PROVIDES THE OPTION
130 REM FOR ADJUSTMENT USING EITHER THE BOWDITCH OR THE MCCAW METHODS
140 INIT
150 SET DEGREES
160 PRINT "ENTER NUMBER OF STATIONS IN TRAVERSE FROM WHICH OBSERVATIONS"
170 PRINT "HAVE BEEN MADE=";
180 INPUT J
190 DIM L(J),B(J),E(J),N(J),D(J),M(J),S(J),U(J),W(J),X(J),Y(J)
200 PRINT "ENTER EASTINGS,NORTHINGS OF INITIAL DATUM TERMINAL";
210 INPUT X1,Y1
220 PRINT "ENTER EASTINGS,NORTHINGS OF FINAL DATUM TERMINAL";
230 INPUT X2,Y2
240 PRINT @4:"     ";"BEARING";"          ";"DISTANCE";"      ";
250 PRINT @4:"DIFF EASTINGS";"      ";"DIFF NORTHINGS"
260 Z1=0
270 Z2=0
280 Z3=0
290 Z4=0
300 Z5=0
310 FOR I=1 TO J-1
320 PRINT "ENTER DEGREES,MINS,SECS OF BEARING OF LEG";I
330 INPUT D(I),M(I),S(I)
340 B(I)=D(I)+M(I)/60+S(I)/3600
350 PRINT "ENTER REDUCED DISTANCE OF LEG";I
360 INPUT L(I)
370 E(I)=L(I)*SIN(B(I))
380 N(I)=L(I)*COS(B(I))
390 Z1=Z1+L(I)
400 Z2=Z2+E(I)
410 Z3=Z3+N(I)
420 Z4=Z4+ABS(E(I))
430 Z5=Z5+ABS(N(I))
440 PRINT @4:B(I);"     ";L(I);"     ";E(I);"     ";N(I)
450 NEXT I
460 PRINT @4:
470 PRINT @4:"SUM PARTIAL EASTINGS=";Z2;"   ";"SUM PARTIAL NORTHINGS=";Z3
480 X3=X2-X1
490 V1=Z2-X3
500 Y3=Y2-Y1
510 V2=Z3-Y3
520 PRINT @4:"SHOULD BE";"          ";X3;"                    ";
530 PRINT @4:"   ";Y3
540 PRINT @4:
550 PRINT @4:"MISCLOSURE EASTINGS=";V1;"   ";"MISCLOSURE NORTHINGS=";V2
560 PRINT @4:
570 PRINT @4:"                    ";"ADJUSTMENT";"           ";
580 PRINT @4:"CORRECTED PARTIAL COORDINATES"
590 PRINT @4:"               ";"EASTINGS";"          ";"NORTHINGS";"    ";
600 PRINT @4:"     ";"EASTINGS";"          ";"NORTHINGS"
610 PRINT "IF MCCAW  ADJUSTMENT REQUIRED PRESS 1 IF BOWDITCH ADJUSTMENT"
620 PRINT "PRESS 2";
630 INPUT K
640 IF K=1 THEN 950
650 PRINT @4:
660 PRINT @4:"BOWDITCH ADJUSTMENT"
670 PRINT @4:
680 U(1)=-(V1/Z1*L(1))
690 W(1)=-(V2/Z1*L(1))
700 E(1)=E(1)+U(1)
710 N(1)=N(1)+W(1)
720 PRINT @4:" A TO 1";"     ";U(1);"   ";W(1);"   ";E(1);"   ";N(1)
730 FOR I=2 TO J-1
740 U(I)=-(V1/Z1*L(I))
750 W(I)=-(V2/Z1*L(I))
```

```
760 E(I)=E(I)+U(I)
770 N(I)=N(I)+W(I)
780 PRINT @4;I-1;" TO ";I;"        ";U(I);"   ";W(I);"   ";E(I);"   ";N(I)
790 NEXT I
800 PRINT @4;
810 PRINT @4;"FINAL COORDINATES"
820 PRINT @4;
830 X(1)=X1+E(1)
840 Y(1)=Y1+N(1)
850 PRINT @4;"FIRST TERMINAL";"    ";X1;"        ";Y1
860 PRINT @4;" 1";"        ";X(1);"   ";Y(1)
870 FOR I=2 TO J-2
880 X(I)=X(I-1)+E(I)
890 Y(I)=Y(I-1)+N(I)
900 PRINT @4;I;"            ";X(I);"   ";Y(I)
910 NEXT I
920 PRINT @4;"FINAL TERMINAL";"    ";X2;"        ";Y2
930 PRINT @4;
940 GO TO 1070
950 U(1)=-(V1/Z4*ABS(E(1)))
960 W(1)=-(V2/Z5*ABS(N(1)))
970 E(1)=E(1)+U(1)
980 N(1)=N(1)+W(1)
990 PRINT @4;
1000 PRINT @4;"MCCAW ADJUSTMENT"
1010 PRINT @4;
1020 PRINT @4;" A TO 1";"    ";U(1);"   ";W(1);"   ";E(1);"   ";N(1)
1030 FOR I=2 TO J-1
1040 U(I)=-(V1/Z4*ABS(E(I)))
1050 W(I)=-(V2/Z5*ABS(N(I)))
1060 GO TO 760
1070 STOP
1080 END
```

4.5.4 The ray trace

A useful procedure commonly adopted when it is not possible to occupy the terminal stations of a traverse or when orientation is not available at terminal stations is to compute the traverse based on an arbitrary bearing allocated to the first leg.

Assuming two such terminals, X and Y (Figure 4.16), and accepting a provisional value for the bearing of XT_1 (let this be α_{XT1}), the partial coordinates for the traverse may first be computed using the reduced distances and provisional bearings, i.e. the bearings based on the assumed initial bearing (α'_{XT1}) and the mean observed angles at each traverse station. The effect of this will be to swing the whole traverse about the datum starting point by an amount equal to ($\alpha_{XT1} - \alpha'_{XT1}$) and to cause the computed coordinates of the final datum point to differ from its datum values. Assuming the two datum points to be *in situ* and their coordinates to be error free, the magnitude of this 'misclosure' will depend on the proximity of α'_{T1} to α_{XT1} and on the presence or otherwise of observational errors in angle and distance measurement.

Denoting the erroneous position of Y by Y' the adjustment requires a comparison to be made between the bearing and distance of XY and XY'. Any difference in bearing represents the *swing* arising from the assumed initial bearing and this difference (θ) must therefore

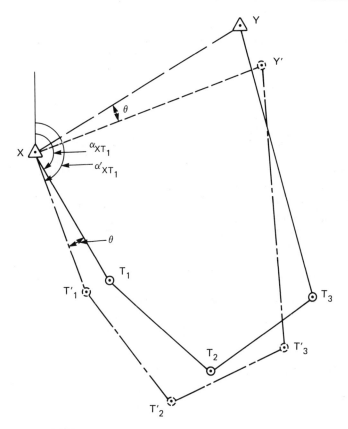

Figure 4.16 Ray trace

be applied to *each* of the provisional bearings. Any small difference in distance is adjusted by applying to each of the reduced distances a *scale factor* determined from the ratio XY/XY' which should be close to unity. The modified bearings and distances may then be used to recompute the traverse which should now close *exactly* on the final datum station Y.

Example 4.22
Using the following data determine coordinates for point T_1 and T_2

	E	(m)	N
A	8901.234		1234.567
B	8637.309		1402.725

	Theodolite horizontal observations (meaned pointings)	Reduced distances (m)
At T$_1$		
To A	99°34'20"	193.456
T$_2$	260 23 00	104.963
At T$_2$		
To T$_1$	80 23 40	
B	09 38 50	146.036

Solution:

Assuming the initial pointing from T$_1$ to A to be the correct bearing then the reverse bearing AT$_1$ = 279°34'20".

	Assumed bearing	Distance	ΔE	ΔN
A				
	279°34'20"	193.456	− 190.762	32.170
T$_1'$				
	260 23 00	104.963	− 103.488	− 17.535
T$_2'$				
	09 38 10	146.036	24.445	143.976
B'				
			ΣΔE = − 269.805	ΣΔN = 158.611

Hence:

$$AB' = 300°27'00" @ 312.973 \text{ m}$$
$$\Delta E_{AB} = -263.925 \quad \Delta N_{AB} = 168.158$$

and:

$$AB' = 302°30'11" @ 312.943$$

Compare with

$$AB' = 300°27'00" \quad 312.973$$
$$\theta = +2°03'11" \quad = 0.030/312.9 = -1/10430 = (\text{Scale Factor})$$

	Adjusted bearing	Adjusted distance	ΔE	ΔN	E	N
A					8901.234	1234.567
	281°37'31"	193.437	− 189.469	38.980		
T$_1$					8711.765	1273.547
	262 26 11	104.953	− 104.040	− 13.815		
T$_2$					8607.725	1259.732
	11 41 21	146.022	29.584	142.994		
B					8637.309	1402.726
			ΣΔE − 263.925	ΣΔN168.159		

Program 4.24 Ray trace computation

```
100 PRINT "PROGRAM 4.24"
110 REM THIS PROGRAM DETERMINES THE COORDINATES OF POINTS INCLUDED IN A
120 REM RAY TRACE WITH AN ASSUMED ORIENTATION FOR THE FIRST LEG
130 INIT
140 SET DEGREES
150 DIM E(10),N(10),B(10),L(10),D(10),M(10),S(10)
160 Z1=0
170 Z2=0
180 PRINT "ENTER EASTINGS, NORTHINGS OF FIRST TERMINAL AND EASTINGS,"
190 PRINT "NORTHINGS OF SECOND TERMINAL";
200 INPUT X1,Y1,X2,Y2
210 PRINT "ENTER NUMBER OF LINES OF RAY TRACE";
220 INPUT Q
230 PRINT @4:"ASSUMED BEARINGS";"            ";"REDUCED DISTANCES"
240 PRINT @4:
250 FOR I=1 TO Q
260 PRINT "ENTER DEGREES,MINS, SECS OF BEARING OF LINE";I;
270 INPUT D(I),M(I),S(I)
280 B(I)=D(I)+M(I)/60+S(I)/3600
290 PRINT "ENTER LENGTH OF LINE";I;
300 INPUT L(I)
310 E(I)=L(I)*SIN(B(I))
320 N(I)=L(I)*COS(B(I))
330 Z1=Z1+E(I)
340 Z2=Z2+N(I)
350 PRINT @4:B(I);"              ";L(I)
360 NEXT I
370 PRINT @4:
380 PRINT @4:"SUM OF PARTIAL EASTINGS=";Z1
390 PRINT @4:"SUM OF PARTIAL NORTHINGS=";Z2
400 X3=X2-X1
410 Y3=Y2-Y1
420 A1=ATN(X3/Y3)
430 A=A1
440 GOSUB 810
450 A1=A
460 A2=ATN(Z1/Z2)
470 A=A2
480 GOSUB 810
490 A2=A
500 L1=SQR(X3^2+Y3^2)
510 L2=SQR(Z1^2+Z2^2)
520 A3=A2-A1
530 L3=L1/L2
540 PRINT @4:"SWING =";-A3
550 PRINT @4:"SCALE FACTOR=";L3
560 Z1=0
570 Z2=0
580 FOR I=1 TO Q
590 B(I)=B(I)-A3
600 L(I)=L(I)*L3
610 E(I)=L(I)*SIN(B(I))
620 N(I)=L(I)*COS(B(I))
630 Z1=Z1+E(I)
640 Z2=Z2+N(I)
650 NEXT I
660 PRINT @4:
670 PRINT @4:"SUM OF CORRECTED PARTIAL EASTINGS =";Z1
680 PRINT @4:"SUM OF CORRECTED PARTIAL NORTHINGS=";Z2
690 PRINT @4:
700 PRINT @4:"FINAL COORDINATES"
710 E(1)=E(1)+X1
720 N(1)=N(1)+Y1
730 PRINT @4: USING 740:E(1),N(1)
740 IMAGE 6D.4D,3X,6D.4D
750 FOR I=2 TO Q
760 E(I)=E(I)+E(I-1)
770 N(I)=N(I)+N(I-1)
780 PRINT @4: USING 740:E(I),N(I)
790 NEXT I
800 GO TO 900
```

```
810 IF A>0 THEN 870
820 IF X3<0 THEN 850
830 A=180+A
840 GO TO 890
850 A=360+A
860 GO TO 890
870 IF X3>0 THEN 890
880 A=180+A
890 RETURN
900 STOP
910 END
```

```
ASSUMED BEARINGS              REDUCED DISTANCES

  279.572222222                     193.456
  260.383333333                     104.963
    9.63611111111                   146.036

SUM OF PARTIAL EASTINGS=-269.805486735
SUM OF PARTIAL NORTHINGS=158.610821273
SWING =2.05293397882
SCALE FACTOR=0.999903649023

SUM OF CORRECTED PARTIAL EASTINGS =-263.925
SUM OF CORRECTED PARTIAL NORTHINGS=168.158

FINAL COORDINATES
   8711.7646      1273.5462
   8607.7249      1259.7313
   8637.3090      1402.7250
```

Chapter 5

Site surveys

Irrespective of size, a fundamental requirement of any construction project is the provision of a topographic map and/or a site plan upon which design proposals may be based. The relative positions and heights of all surface features existing on the site, natural and man-made, must be predetermined and represented either graphically, in the form of a map or plan, or numerically, in the form of a digital model, to an order of accuracy which satisfies the requirements and conditions stipulated in the project specification.

An increasing amount of site surveying today is being undertaken using instruments which provide the facility not only for the precise measurement of angles but which also incorporate, either by integration in the instrument's design or by use of a theodolite 'add-on' facility, the means for the rapid and accurate determination of distance and difference in altitude by electromagnetic measurement. In addition, some instruments are equipped for the automatic collection and storage of such vector data which may in turn be transferred via computer to automatic plotting devices thereby reducing considerably the time, labour, and, in particular circumstances, the cost, of the acquisition, processing and representation of site data.

However, such facilities are not always available and many surveyors and engineers must continue to make use of more conventional techniques for site surveying.

5.1 Tacheometric surveying

Tacheometric measurement provides a convenient, indirect method for obtaining horizontal distances and differences in elevation between an instrument station and the position of a vertically held graduated staff. If the instrument station is connected to a control framework and if the directions to staff positions are also included in the observations then the relative position and height of each staff position may be determined.

In modern theodolites the optics of the telescope and the

separation of the stadia hairs are so arranged that the following two
basic tacheometric relationships hold (Figure 5.1):

$$D = 100s \cos^2 V \tag{5.1}$$
$$h = 50s \sin 2V \tag{5.2}$$

where D = the horizontal distance between the instrument station and
the staff position, h = the difference in height between the theodolite
trunnion axis and the horizontal cross hair reading on the vertical
staff, s = the difference between the upper (u) and lower (l) stadia
readings on the staff (the staff intercept), V = the angle of elevation
/depression.

Hence, if the altitude of the instrument station (A_1) is known, the
height of the staff position (A_2) may be determined from:

$$H = h + i - m$$
$$A_2 = A_1 + H$$

where i = the height of the theodolite trunnion axis above the ground
station, and m = the horizontal cross hair reading on the staff.

In the past the reduction of tacheometric observations tended to be
time consuming and laborious. The use of tacheometric tables
providing values for D and H at constant intervals for V and s helped
to facilitate the reduction process but necessitated the setting of the
theodolite telescope at a vertical angle corresponding to some
multiple of the V interval. The use of a simple computer program
using the above fundamental formulae with random values for V
greatly facilitates the reduction of large quantities of tacheometric
field data.

Figure 5.1 Tacheometry: basic principle

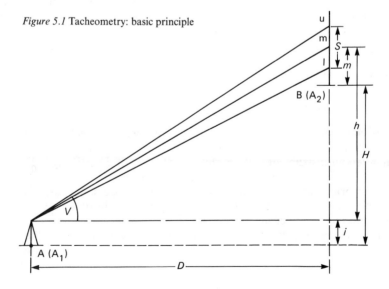

Table 5.1 A typical tacheometric reduction sheet

Inst Station: A

Altitude of instrument station = A_1 = 98.27 m
Height of inst above ground mark = i = 1.36 m

1 Staff station	2 Horizontal circle	3 Stadia u m l $s = (u - l)$	4 Vertical angle (V) E = Elevation D = Depression	5 D	6 h	7 $h - m$	8 H(m)	9 A(m)
B	0° 00'	1.404 0.992 0.581 ———— 0.823	2°40'E	82.1	+ 3.82	+ 2.83	4.19	102.46
118	240 07	3.346 2.794 2.244 ———— 1.102	6°40'D	108.7	− 12.70	− 15.49	− 14.13	84.14

Table 5.1 illustrates a typical tacheometric reduction sheet, the contents of columns 1–4 being field data and 5–9 reduced data, that is:

Column 5
$$D_B = 100\,(1.404 - 0.581)\cos^2 2°40' = 82.1\,\text{m}$$
Column 6
$$h_B = 50(1.404 - 0.581)\sin 2(2°40') = +3.82\,\text{m}$$
Column 7
$$h_B - m_B = 3.82 - 0.992 = +2.83\,\text{m}$$
Column 8
$$H_B = 2.83 + 1.36 = +4.19\,\text{m}$$
Column 9
$$A_B = 98.27 + 4.19 = 102.46\,\text{m}$$

Program 5.1 Tacheometric reduction

```
TACHEOMETRIC REDUCTION

HEIGHT OF INSTRUMENT=1.36    ALTITUDE OF STATION=98.27

                  STADIAS          VERT ANGLE    REDUCED DISTANCE   REDUCED ALTITUDE
STATION 1  1.404  0.992  0.581    2  40  0           82.1219           102.4629
STATION 2  3.346  2.794  2.244   -6  40  0          108.7148            84.1291
```

```
100 PRINT "PROGRAM 5.1"
110 INIT
120 SET DEGREES
130 PRINT @4:"TACHEOMETRIC REDUCTION"
140 PRINT @4:
150 PRINT @4:
160 PRINT "ENTER NUMBER OF LINES TO BE REDUCED";
170 INPUT N
180 DIM T(N),C(N),B(N),V(N),L(N),H(N)
190 PRINT "HEIGHT OF INSTRUMENT=";
200 INPUT I
210 PRINT "ALTITUDE OF STATION=";
220 INPUT A
230 PRINT @4:"HEIGHT OF INSTRUMENT=";I;"     ";"ALTITUDE OF STATION=";A
240 PRINT @4:
250 PRINT @4:"                   ";"STADIAS";"        ";"VERT ANGLE";
260 PRINT @4:"   ";"REDUCED DISTANCE";"   ";"REDUCED ALTITUDE"
270 FOR J=1 TO N
280 PRINT "ENTER TOP, CENTRE,BOTTOM STADIA READINGS=";
290 INPUT T(J),C(J),B(J)
300 V(J)=0
310 PRINT "ENTER SIGN,DEGREES,MINS,SECS OF VERTICAL ANGLE=";
320 INPUT D,M,S
330 V(J)=SGN(D)*(ABS(D)+M/60+S/3600)
340 L(J)=(T(J)-B(J))*100*COS(V(J))^2
350 H(J)=(T(J)-B(J))*100*SIN(V(J))*COS(V(J))-C(J)+A+I
360 PRINT @4: USING 380:J,T(J),C(J),B(J),D,M,S,L(J),H(J)
380 IMAGE"STATION ",D,3(2X,D.3D),X,3D,2(X,2D),2(6X,4D.4D)
400 NEXT J
410 PRINT @4:
420 PRINT @4:
430 STOP
440 END
```

5.2 Levelling

If specifications require the accurate determination of heights over a site, the use of levelling instruments designed specifically for the direct determination of height differences is necessary.

In Figure 5.2, the height difference between a benchmark A of

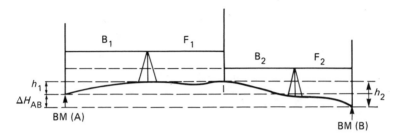

Figure 5.2 Levelling: principle of height determination

known height (H_A) above datum and N, any point whose height is required, may be expressed as:

$$\Delta H_{AN} = (B_1 - F_1) + (B_2 - F_2) + \ldots + (B_{n-1} - F_{n-1}) + (B_n - F_n)$$
$$= h_1 + h_2 + \ldots + h_{n-1} + h_n \qquad (5.3)$$

where $B_1 \ldots B_n$ and $F_1 \ldots F_n$ are the respective backsight and foresight observations made to n successive positions of the levelling staff and $h_1 \ldots h_n$ are the differences in height between successive staff positions in the circuit. The height of N is therefore:

$$H_N = H_A + \Delta H_{AN}$$

To provide a check and to demonstrate the accuracy of the results the levels must be closed, either onto the initial benchmark or onto a second benchmark of known height above datum (Figure 5.3).

The processing of the field data is normally carried out by tabulating the observations and reducing them in one of two ways:

1. The 'rise and fall' method provides a check on the arithmetic reduction of the observed data (including that relating to 'intermediate' points) using the condition that:

$$\Sigma B - \Sigma F = \Sigma \text{RISE} + \Sigma \text{FALL} = H'_B - H_A$$

where H'_B is the derived height of B.

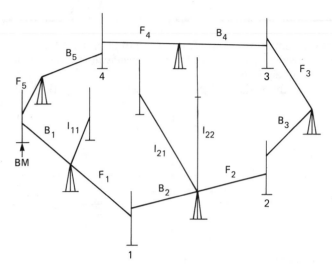

Figure 5.3 Levelling: closed circuit with back, intermediate and fore sights

2. The 'height of collimation' method also provides a check on the arithmetic reduction in that:

$$\Sigma B - \Sigma F = H'_B - H_A$$

But, unlike the rise and fall reduction, the reduced heights of any intermediate points are unchecked.

In both methods the accuracy of the levelling may be demonstrated by determining the magnitude of the misclosure (dH):

$$dH = H'_B - H_B$$

where H_B is the datum height of the final terminal benchmark.

The misclosure may then be distributed between the observed height differences according to the number of instrument change points, an intermediate point receiving the same correction as its associated change point.

Examples 5.2 (Table 5.2) and 5.3 (Table 5.3) illustrate respectively the two reduction methods described above.

Table 5.2 Example 5.2, the 'rise and fall' method of reduction

Back sight (B)	Inter sight (I)	Fore sight (F)	Rise	Fall	Derived altitude (H')	Adjustment (mm)	Reduced altitude (H)	
1.290	0.488						110.941	BM(A)
	0.854		0.802		111.743		111.743	
				0.366	111.377		111.377	
2.829		0.448	0.406		111.783		111.783	TP1
	2.011		0.818		112.601	+3	112.604	
	0.945		1.066		113.667	+3	113.670	
	0.756		0.189		113.856	+3	113.859	
1.670		0.994		0.238	113.618	+3	113.621	TP2
	2.652			0.982	112.636	+5	112.641	
	0.640		2.012		114.648	+5	114.653	
	3.152			2.512	112.136	+5	112.141	
		2.371	0.781		112.917	+5	112.922	BM(B)
						(= dH)		
Σ5.789		3.813	6.074	4.098	110.941			
Σ3.813			4.098		1.976			
Δ1.976			1.976					

116

Table 5.3 Example 5.3, the 'height of collimation' method of reduction

Back sight (B)	Inter sight (I)	Fore sight (F)	Height of collimation	Derived altitude (H')	Adjustment (mm)	Reduced altitude (H)	Remarks
0.172			18.542			18.370	BM(A)
0.217	−1.155	0.908	17.851	17.634		17.634	
				19.006	−2	19.004	Inverted staff to underside of bridge
0.983		0.375	18.459	17.476	−2	17.474	
	0.867			17.592	−3	17.589	
	0.552			17.907	−3	17.904	
		0.183		18.276	−3	18.273	BM(B)
					(= dH)		
Σ 1.372		1.466		18.370			
Σ		1.372					
Δ		0.094		0.094			

Program 5.2 Levelling reduction (rise and fall)

BS	IS	FS	R	F	H
1.290					110.941
	0.488		0.802		111.743
	0.854			0.366	111.377
2.829		0.448	0.406		111.783
	2.011		0.818		112.601
	0.945		1.066		113.667
	0.756		0.189		113.856
1.670		0.994		0.238	113.618
	2.652			0.982	112.636
	0.640		2.012		114.648
	3.152			2.512	112.136
		2.371	0.781		112.917

SUM	5.789	3.813			112.917
DIFF	1.976		1.976		1.976
MISCLOSURE					−0.005

```
100 PRINT "PROGRAM 5.2"
110 REM THIS PROGRAM REDUCES FIELD LEVELLING OBSERVATIONS USING THE
120 REM 'RISE AND FALL' REDUCTION METHOD. (EXAMPLE 5.2)
130 INIT
140 PRINT "ENTER NUMBER OF INSTRUMENT STATIONS"
150 INPUT N
160 DIM B(N),F(N),D(10),I(20),H(20)
170 PRINT "ENTER ALTITUDES OF TERMINAL BENCHMARKS BM1 AND BM2"
180 INPUT A1,A2
190 Z1=0
200 Z2=0
210 Z3=0
220 PRINT @4:"          ";"BS";"          ";"IS";"          ";"FS";"          ";"R";
230 PRINT @4:"          ";"F";"          ";"H"
240 PRINT @4:
250 PRINT "ENTER BACKSIGHT FROM STATION 1"
260 INPUT B(1)
270 Z1=B(1)
280 PRINT @4: USING 290:B(1),A1
290 IMAGE 8X,D.3D,35X,4D.3D
300 PRINT "ENTER NUMBER OF INTERMEDIATE SIGHTS FROM STATION 1"
310 INPUT M
320 PRINT "ENTER INTERMEDIATE SIGHT 1 FROM STATION 1"
330 INPUT I(1)
340 D(1)=B(1)-I(1)
350 Z3=D(1)
360 H(1)=A1+D(1)
370 IF D(1)<0 THEN 400
380 PRINT @4: USING 790:I(1),ABS(D(1)),H(1)
390 GO TO 410
400 PRINT @4: USING 760:I(1),ABS(D(1)),H(1)
410 FOR J=2 TO M
420 PRINT "ENTER INTERMEDIATE SIGHT";J;"FROM STATION";1
430 INPUT I(J)
440 D(J)=I(J-1)-I(J)
450 Z3=Z3+D(J)
460 H(J)=H(J-1)+D(J)
470 IF D(J)<0 THEN 500
480 PRINT @4: USING 790:I(J),ABS(D(J)),H(J)
490 GO TO 510
500 PRINT @4: USING 760:I(J),ABS(D(J)),H(J)
510 NEXT J
520 PRINT "ENTER FORESIGHT FROM STATION 1"
530 INPUT F(1)
540 Z2=F(1)
550 D(M+1)=I(M)-F(1)
```

```
560 Z3=Z3+D(M+1)
570 H(M+1)=H(M)+D(M+1)
580 PRINT "ENTER BACKSIGHT FROM STATION 2"
590 INPUT B(2)
600 Z1=Z1+B(2)
610 IF D(M+1)<0 THEN 640
620 PRINT @4: USING 1050:B(2),F(1),D(M+1),H(M+1)
630 GO TO 650
640 PRINT @4: USING 1080:B(2),F(1),ABS(D(M+1)),H(M+1)
650 P=H(M+1)
660 FOR K=2 TO N
670 PRINT "ENTER NUMBER OF INTERMEDIATE SIGHTS FROM STATION";K
680 INPUT M
690 PRINT "ENTER INTERMEDIATE SIGHT";1;"FROM STATION";K
700 INPUT I(1)
710 D(1)=B(K)-I(1)
720 Z3=Z3+D(1)
730 H(1)=P+D(1)
740 IF D(1)<0 THEN 780
750 PRINT @4: USING 790:I(1),D(1),H(1)
760 IMAGE 16X,D.3D,19X,D.3D,3X,4D.3D
770 GO TO 800
780 PRINT @4: USING 760:I(1),ABS(D(1)),H(1)
790 IMAGE 16X,D.3D,11X,D.3D,11X,4D.3D
800 FOR J=2 TO M
810 PRINT "ENTER INTERMEDIATE SIGHT";J;"FROM STATION";K
820 INPUT I(J)
830 D(J)=I(J-1)-I(J)
840 Z3=Z3+D(J)
850 H(J)=H(J-1)+D(J)
860 IF D(J)<0 THEN 890
870 PRINT @4: USING 790:I(J),D(J),H(J)
880 GO TO 900
890 PRINT @4: USING 760:I(J),ABS(D(J)),H(J)
900 NEXT J
910 PRINT "ENTER FORESIGHT FROM STATION";K
920 INPUT F(K)
930 Z2=Z2+F(K)
940 D(M+1)=I(M)-F(K)
950 Z3=Z3+D(M+1)
960 H(M+1)=H(M)+D(M+1)
970 IF K=N THEN 1110
980 PRINT "ENTER BACKSIGHT FROM STATION";K+1
990 INPUT B(K+1)
1000 Z1=Z1+B(K+1)
1010 D(M+1)=I(M)-F(K)
1020 H(M+1)=H(M)+D(M+1)
1030 IF D(M+1)<0 THEN 1070
1040 PRINT @4: USING 1050:B(K+1),F(K),D(M+1),H(M+1)
1050 IMAGE 8X,D.3D,11X,D.3D,3X,D.3D,11X,4D.3D,5X
1060 GO TO 1090
1070 PRINT @4: USING 1080:B(K+1),F(K),ABS(D(M+1)),H(M+1)
1080 IMAGE 8X,D.3D,11X,D.3D,11X,D.3D,3X,4D.3D,5X
1090 P=H(M+1)
1100 NEXT K
1110 IF D(M+1)<0 THEN 1150
1120 PRINT @4: USING 1130:F(K),D(M+1),H(M+1)
1130 IMAGE 24X,D.3D,3X,D.3D,11X,4D.3D
1140 GO TO 1170
1150 PRINT @4: USING 1160:F(K),ABS(D(M+1)),H(M+1)
1160 IMAGE 24X,D.3D,11X,D.3D,3X,4D.3D
1170 PRINT @4:
1180 PRINT @4:
1190 PRINT @4: USING 1210:Z1,Z2,H(M+1)
1210 IMAGE "SUM",4X,D.3D,10X,D.3D,20X,4D.3D
1220 Y=H(M+1)-A1
1230 X=Z1-Z2
1240 PRINT @4:
1250 PRINT @4: USING 1260:X,ABS(Z3),Y
1260 IMAGE "DIFF",12X,D.3D,16X,D.3D,6X,4D.3D
1270 V=H(M+1)-A2
1280 PRINT @4: USING 1290:V
```

```
1290 IMAGE"MISCLOSURE",40X,2D.3D
1300 STOP
1310 END
```

Program 5.3 Levelling reduction (height of collimation)

BS	IS	FS	C	H
0.172			18.542	18.370
0.217		0.908	17.851	17.634
	-1.155			19.006
0.983		0.375	18.459	17.476
	0.867			17.592
	0.552			17.907
		0.183		18.276

SUM	1.372		1.466	
DIFF		-0.094		-0.094
MISCLOSURE				0.000

```
100 PRINT "PROGRAM 5.3"
110 REM THIS PROGRAM REDUCES FIELD LEVELLING OBSERVATIONS USING THE
120 REM 'HEIGHT OF COLLIMATION' REDUCTION METHOD. (EXAMPLE 5.3)
130 INIT
140 PRINT "ENTER NUMBER OF INSTRUMENT STATIONS"
150 INPUT N
160 DIM B(N),F(N),C(N),I(20),H(20)
170 Z1=0
180 Z2=0
190 Z3=0
200 PRINT "ENTER ALTITUDES OF TERMINAL BENCHMARKS BM1 AND BM2"
210 INPUT A1,A2
220 PRINT @4: USING 230:
230 IMAGE10X,"BS",6X,"IS",6X,"FS",6X,"C",8X,"H"
240 PRINT @4:
250 PRINT "ENTER BACKSIGHT FROM STATION 1"
260 INPUT B(1)
270 Z1=B(1)
280 C(1)=A1+B(1)
290 PRINT @4: USING 300:B(1),C(1),A1
300 IMAGE 8X,D.3D,19X,2D.3D,2X,4D.3D
310 PRINT "ENTER FORE SIGHT FROM STATION 1"
320 INPUT F(1)
330 Z2=F(1)
340 H(1)=C(1)-F(1)
350 PRINT "ENTER BACKSIGHT FROM STATION 2"
360 INPUT B(2)
370 C(2)=H(1)+B(2)
380 Z1=Z1+B(2)
390 PRINT @4: USING 400:B(2),F(1),C(2),H(1)
400 IMAGE 8X,D.3D,11X,D.3D,3X,2D.3D,2X,4D.3D
410 FOR K=2 TO N
420 PRINT "ENTER NUMBER OF INTERMEDIATE SIGHTS FROM STATION";K
430 INPUT M
440 FOR J=1 TO M
450 PRINT "ENTER INTERMEDIATE SIGHT";J;"FROM STATION";K
460 INPUT I(J)
470 H(J)=C(K)-I(J)
480 PRINT @4: USING 500:I(J),H(J)
490 NEXT J
500 IMAGE 16X,2D.3D,18X,4D.3D
510 FOR J=2 TO M
520 PRINT "ENTER FORE SIGHT FROM STATION";K
530 INPUT F(K)
540 Z2=Z2+F(K)
550 H(K)=C(K)-F(K)
```

```
560 IF K=N THEN 640
570 PRINT "ENTER BACKSIGHT FROM STATION";K+1
580 INPUT B(K+1)
590 C(K+1)=H(M+1)+B(K+1)
600 Z1=Z1+B(K+1)
610 PRINT @4: USING 620:B(K+1),F(K),C(K+1),H(M+1)
620 IMAGE8X,D.3D,11X,D.3D,3X,2D.3D,2X,4D.3D
630 NEXT K
640 PRINT @4: USING 650:F(K),H(M+1)
650 IMAGE 24X,D.3D,11X,4D.3D
660 PRINT @4:
670 PRINT @4:
680 PRINT @4: USING 690:Z1,Z2
690 IMAGE "SUM",5X,D.3D,10X,2D.3D
700 X=Z1-Z2
710 Y=H(K)-A1
720 PRINT @4: USING 730:X,Y
730 IMAGE "DIFF",12X,2D.3D,18X,4D.3D
740 V=H(K)-A2
750 PRINT @4: USING 760:V
760 IMAGE "MISCLOSURE",32X,2D.3D
770 STOP
780 END
```

Chapter 6

Earthworks

6.1 Determination of area and volume

With the increasing cost of land and materials, it is vital that the engineer is able to make assessments of relevant quantities involved in any particular project in accordance with specified accuracies. Estimation of areas and volumes is fundamental to the majority of engineering projects especially the implementation of highway and tunnelling programmes, dam construction and the determination of reservoir capacity.

6.1.1 Area determination

6.1.1.1 *Rectilinear boundaries*

The following formulae are commonly used for the determination of the area of straight-sided figures capable of being divided into a series of triangles (Figure 6.1):

$$\text{Area } (A) = [s(s-a)(s-b)(s-c)]^{\frac{1}{2}} \tag{6.1}$$

where $s = \frac{1}{2}(a+b+c)$.

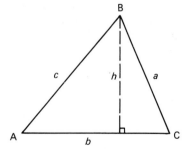 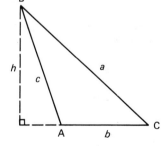

Figure 6.1 Area determination: triangle

Also:

$$A = \tfrac{1}{2}ab \sin C = \tfrac{1}{2}ac \sin B = \tfrac{1}{2}bc \sin A \qquad (6.2)$$

$$A = \frac{\tfrac{1}{2}c^2}{\cot A + \cot B} = \frac{\tfrac{1}{2}a^2}{\cot B + \cot C} = \frac{\tfrac{1}{2}b^2}{\cot A + \cot C} \qquad (6.3)$$

$$A = \tfrac{1}{2}hb \qquad (6.4)$$

6.1.1.2 Irregular boundaries

Where one or more of the boundaries are curvilinear, the figure may be subdivided into a series of triangles, the area of which may be calculated using one of the formulae in Section 6.1.1.1 (Figure 6.2).

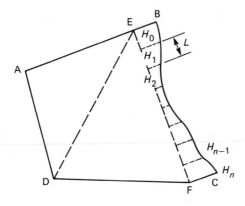

Figure 6.2 Area determination: curvilinear boundary

One of the rectilinear boundaries thus formed may then be adopted as a base line from which offsets to a curvilinear boundary may be determined and either the trapezoidal rule or Simpson's rule applied to calculate the area of the curvilinear section.

Trapezoidal rule:

$$A = L\left[\frac{(H_0 + H_n)}{2} + \sum_{i=1}^{(n-1)} H_i\right] \qquad (6.5)$$

where $H_0, H_1, H_2 \ldots H_n$ are offsets from the rectilinear base line (odd or even in number) and L is the uniform distance between the offsets.

Simpson's rule:

$$A = \frac{L}{3}[(H_0 + H_n) + 2(H_2 + H_4 + \ldots H_{n-1}) + 4(H_1 + H_3 \ldots H_{n-2})]$$

$$= \frac{L}{3}[(H_0 + H_n) + 2\Sigma H_{even} + 4\Sigma H_{odd})] \tag{6.6}$$

where n is an odd number.

Example 6.1
Given the following data determine the area of property ABCFD (Figure 6.2) to the nearest 0.1 ha.

Distance	Offsets at 200 m intervals from E (m)	
AD = 1114.5 m	H_0	231
DF = 1420.2 m	H_1	193
FC = 281.1 m	H_2	111
AE = 1185.4 m	H_3	73
EB = 212.6 m	H_4	88
DE = 1650.0 m	H_5	121
EF = 1601.3 m	H_6	144
	H_7	197
	H_8	281

Solution:

A = area of (AED + DEF) + curvilinear area (EBCF).

Using Equation (6.1), for AED:
$S = \frac{1}{2}(1185.4 + 1114.5 + 1650.0) = 1974.95$

Area = $[1974.95(1974.95 - 1185.4)(1974.95 - 1114.5)(1974.95 - 1650.0)]^{1/2}$
= 660 296 sq m

For DEF:

$S = \frac{1}{2}(1650.0 + 1601.3 + 1420.2) = 2335.77$
Area = $[2335.77(2335.77 - 1650.0)(2335.77 - 1601.3)(2335.77 - 1420.2)]^{1/2}$
= 1 037 856 sq m

Using Simpson's rule, the curvilinear area is:

$$A = \frac{200}{3}[(231 + 281) + 2(193 + 73 + 121 + 197) + 4(111 + 88 + 144)]$$

= 203 467
Total area = 660 296 + 103 785 6 + 203 467 = 190 161 9 m^2
= 190.2 ha

Program 6.1 Area determination (rectilinear/curvilinear)

```
        1185.4   1114.5   1650.0     66.03
        1650.0   1601.3   1420.2    103.78

TOTAL RECTILINEAR AREA              169.81


       EVEN OFFSETS       ODD OFFSETS

          231
          111
           88
          144
          281
                             193
                              73
                             121
                             197


TOTAL CURVILINEAR AREA              20.35

TOTAL AREA                          190.2
```

```
100 PRINT "PROGRAM 6.1"
110 REM THIS PROGRAM CALCULATES THE AREA OF A FIGURE COMPRISING
120 REM COMPONENTS BOUNDED BY RECTILINEAR AND CURVILINEAR BOUNDARIES
130 REM (EXAMPLE 6.1)
140 INIT
150 PRINT "ENTER NUMBER OF TRIANGULAR UNITS"
160 INPUT M
170 DIM A(M)
180 Z=0
190 FOR I=1 TO M
200 DIM S(M)
210 PRINT "ENTER LENGTHS OF SIDES OF TRIANGLE";I
220 INPUT L1,L2,L3
230 S(I)=(L1+L2+L3)*0.5
240 A(I)=SQR(S(I)*(S(I)-L1)*(S(I)-L2)*(S(I)-L3))
250 A(I)=A(I)/10000
260 PRINT @4: USING 270:L1,L2,L3,A(I)
270 IMAGE 8X,3(4D.D,2X),4D.2D
280 Z=Z+A(I)
290 NEXT I
300 PRINT @4:
310 PRINT @4: USING 320:Z
320 IMAGE "TOTAL RECTILINEAR AREA",9X,5D.2D
330 PRINT "ENTER NUMBER OF OFFSETS TO CURVILINEAR BOUNDARY"
340 INPUT N
350 DIM H(N)
360 PRINT "ENTER OFFSETS 1 TO ";N
370 INPUT H(1),H(2),H(3),H(4),H(5),H(6),H(7),H(8),H(9)
380 PRINT "ENTER INTERVAL BETWEEN OFFSETS"
390 INPUT L
400 PRINT @4:
410 PRINT @4:
420 PRINT @4: USING 430:
430 IMAGE 6X,"EVEN OFFSETS",6X,"ODD OFFSETS"
440 PRINT @4:
450 V=0
460 PRINT @4: USING 470:H(1)
470 IMAGE 10X,3D
480 FOR J=3 TO N-2 STEP 2
490 V=V+H(J)
500 PRINT @4: USING 470:H(J)
510 NEXT J
520 PRINT @4: USING 470:H(N)
530 W=0
540 FOR K=2 TO N-1 STEP 2
```

```
550 W=W+H(K)
560 PRINT @4: USING 570:H(K)
570 IMAGE 28X,3D
580 NEXT K
590 C=L/3*(H(1)+H(N)+2*W+4*V)
600 C=C/10000
610 PRINT @4:
620 PRINT @4:
630 PRINT @4: USING 640:C
640 IMAGE "TOTAL CURVILINEAR AREA",9X,5D.2D
650 T=Z+C
660 PRINT @4:
670 PRINT @4: USING 680:T
680 IMAGE "TOTAL AREA",21X,5D.D
690 STOP
700 END
```

6.1.1.3 Area from coordinates

Referring again to Figure 6.1, if the boundaries of the figure are rectilinear and the coordinates of the turning points are known $(E_1N_1, E_2N_2, E_3N_3 \ldots E_{n-1}N_{n-1}, E_nN_n)$ then:

$$A = \tfrac{1}{2}[(E_1(N_2 - N_n) + E_2(N_3 - N_1) + \ldots E_n(N_1 - N_{n-1})] \qquad (6.7a)$$

Also:

$$A = \tfrac{1}{2}[(N_1(E_2 - E_n) + N_2(E_3 - E_1) \ldots N_n(E_1 - E_{n-1})] \qquad (6.7b)$$

Example 6.2
Referring again to Figure 6.2 determine the area of the rectilinear figure AEFD given the following coordinates

	E	(m)	N
A	10 507.11		36 742.90
E	11 693.50		36 793.11
F	11 879.47		35 202.65
D	10 524.63		35 628.54

Solution

Area $= \tfrac{1}{2}[10\ 524.63(36\ 742.90 - 35\ 202.65)$
$+ 10\ 507.11(36\ 793.11 - 35\ 628.54)$
$+ 11\ 693.50(35\ 202.65 - 36\ 742.90)$
$+ 11\ 879.47(35\ 628.54 - 36\ 793.11)]$
$= 1\ 699\ 280.7\ m^2$

and:

Area $= \tfrac{1}{2}[35\ 628.54(10\ 507.11 - 11\ 879.47)$
$+ 36\ 742.90(11\ 693.50 - 10\ 524.63)$
$+ 36\ 793.11(11\ 879.47 - 10\ 507.11)$
$+ 35\ 202.65(10\ 524.63 - 11\ 693.50)]$
$= 1\ 699\ 280.7\ m^2$ (check)

Program 6.2 Area determination (coordinates)

EASTINGS	NORTHINGS
10507.110	36742.900
11693.500	36793.110
11879.470	35202.650
10524.630	35628.540

```
AREA=    1699280.65
AREA=   -1699280.65(CHECK)
```

```
100 PRINT "PROGRAM 6.2"
110 REM THIS PROGRAM DETERMINES THE AREA OF A POLYGON GIVEN THE
120 REM COORDINATES OF THE BOUNDARY TURNING POINTS. COORDINATES ARE
130 REM ENTERED FOR EACH POINT IN TURN, CLOCKWISE AROUND THE FIGURE
140 INIT
150 PRINT "ENTER NUMBER OF POINTS =";
160 INPUT M
170 DIM E(M),N(M)
180 PRINT @4: USING 190:
190 IMAGE 10X,"EASTINGS",11X,"NORTHINGS"
200 PRINT @4:
210 FOR I=1 TO M
220 PRINT "ENTER EASTINGS, NORTHINGS OF POINT";I;
230 INPUT E(I),N(I)
240 PRINT @4: USING 250:E(I),N(I)
250 IMAGE8X,7D.3D,8X,7D.3D
260 NEXT I
270 A=0
280 E9=E(2)-E(M)
290 B=E9*N(1)
300 A=A+B
310 FOR I=2 TO M-1
320 E9=E(I+1)-E(I-1)
330 B=E9*N(I)
340 A=A+B
350 NEXT I
360 E9=E(1)-E(M-1)
370 B=E9*N(M)
380 A=A+B
390 PRINT @4:
400 PRINT @4: USING 410:A/2
410 IMAGE "AREA=",10D.2D
420 A=0
430 N9=N(2)-N(M)
440 B=N9*E(1)
450 A=A+B
460 FOR I=2 TO M-1
470 N9=N(I+1)-N(I-1)
480 B=N9*E(I)
490 A=A+B
500 NEXT I
510 N9=N(1)-N(M-1)
520 B=N9*E(M)
530 A=A+B
540 PRINT @4: USING 550:A/2
550 IMAGE "AREA=",10D.2D,"(CHECK)"
560 STOP
570 END
```

6.1.1.4 Areas of cross-sections

Volumetric calculations frequently involve cross-sections taken at right angles to the centre line of some feature such as a road or railway alignment.

One, two or three-level section
(a) The area of cross-sections of the form shown in Figure 6.3 may be determined as follows:

$$A = \tfrac{1}{2} W \left(H + \frac{B}{2s} \right) - \frac{B^2}{4s} \tag{6.8}$$

where $W = W_L + W_R$ and W_L, $W_R =$ the side widths, $H =$ the difference in height on the centre line between the formation level and the level of the existing terrain, $B =$ the formation width, and $1 : s =$ the gradient of the side slopes.

For a one-level section (Figure 6.3(a)):

$$W_L = W_R = \tfrac{1}{2} B + sH \tag{6.9}$$

For a two-level section (Figure 6.3(b)):
The transverse slope $(1 : k)$ is constant therefore:

$$W_L = \frac{ks}{k \pm s} \left(H + \frac{B}{2s} \right) \quad \text{and} \quad W_R = \frac{ks}{k \pm s} \left(H + \frac{B}{2s} \right) \tag{6.10}$$

For a three-level section (Figure 6.3(c)):
The transverse slope is not constant, i.e. $k_L \neq k_R$, therefore:

$$W_L = \frac{k_L s}{k_L \pm s} \left(H + \frac{B}{2s} \right) \quad \text{and} \quad W_R = \frac{k_R s}{k_R \pm s} \left(H + \frac{B}{2s} \right) \tag{6.11}$$

Example 6.3
Determine the area of the cross-section shown in Figure 6.3(c) given the following data:

Formation width $= 20$ m
Formation height above terrain at centre line $= 10$ m
Side slopes $= 1$ in 2
Transverse gradients of terrain $=$ Left $- 1$ in 20
Right $+ 1$ in 10

Solution:
From Equation (6.11):

$$W_L = \left(\frac{20 \times 2}{20 - 2} \right) \left(10 + \frac{20}{2 \times 2} \right) = 33.3 \text{ m}$$

$$W_R = \left(\frac{10 \times 2}{10 + 2} \right) \left(10 + \frac{20}{2 \times 2} \right) = 25.0 \text{ m}$$

$$W_L + W_R = 58.3 \text{ m}$$

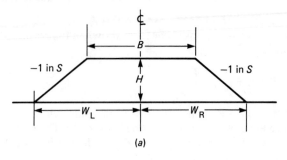

(a)

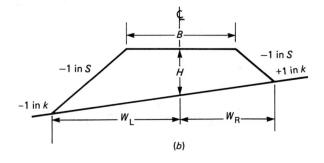

(b)

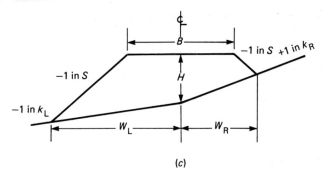

(c)

Figure 6.3 Simple earthwork cross-sections

$$A = \tfrac{1}{2}(58.3)\left(10 + \frac{20}{2 \times 2}\right) - \left(\frac{20^2}{4 \times 2}\right)$$

$$= 387.3 \, \text{m}^2$$

Program 6.3 Cross-section area (1, 2, 3 level)

```
FORMATION WIDTH 20   FORMATION HEIGHT 10.00
SIDE SLOPES 1 IN 2
TRANSVERSE GROUND SLOPE: LEFT 1 IN -20 RIGHT 1 IN   10

LEFT SIDE WIDTH 33.33   RIGHT SIDE WIDTH 25.00

AREA   387.5


100 PRINT "PROGRAM 6.3"
110 REM THIS PROGRAM DETERMINES THE AREA OF A 1-,2-,OR 3-LEVEL ROAD
120 REM CROSS-SECTION (EXAMPLE 6.3)
130 INIT
140 PRINT "ENTER FORMATION WIDTH"
150 INPUT B
160 PRINT "ENTER HORIZONTAL COMPONENT OF GRADIENT OF SIDE SLOPES"
170 INPUT S
180 PRINT "ENTER HEIGHT OF FORMATION LEVEL ABOVE(+), OR BELOW(-)"
190 PRINT "GROUND LEVEL AT CENTRE LINE"
200 INPUT H
210 PRINT "ENTER HORIZONTAL COMPONENT OF GRADIENT OF LEFT TRANSVERSE";
220 PRINT "GROUND SLOPE"
230 INPUT K1
240 PRINT "ENTER HORIZONTAL COMPONENT OF GRADIENT OF RIGHT TRANSVERSE";
250 PRINT "GROUND SLOPE"
260 INPUT K2
270 W1=K1*S/(K1+S)*(H+B/(2*S))
280 W2=K2*S/(K2+S)*(H+B/(2*S))
290 A=(W1+W2)/2*(H+B/(2*S))-B^2/(4*S)
300 PRINT @4: USING 310:B,H
310 IMAGE 6X,"FORMATION WIDTH",3D,2X,"FORMATION HEIGHT",3D.2D
320 PRINT @4: USING 330:S
330 IMAGE 6X,"SIDE SLOPES 1 IN",2D
340 PRINT @4: USING 350:K1,K2
350 IMAGE 6X,"TRANSVERSE GROUND SLOPE: LEFT 1 IN ",3D,X,"RIGHT 1 IN ",3D
360 PRINT @4:
370 PRINT @4: USING 380:W1,W2
380 IMAGE 6X,"LEFT SIDE WIDTH",3D.2D,2X,"RIGHT SIDE WIDTH",3D.2D
390 PRINT @4:
400 PRINT @4: USING 410:A
410 IMAGE 6X,"AREA",5D.D
420 STOP
430 END
```

(b) *Side-hill section (Figure 6.4)*

Assuming different side slopes (s_L and s_R) and that the terrain surface cuts the formation surface eccentric to the centre line:

$$W_L = \frac{ks_L}{k \pm s_L}\left(\frac{B}{2s_L} \pm H\right) \quad \text{and} \quad W_R = \frac{ks_R}{k \pm s_R}\left(\frac{B}{2s_R} \pm H\right) \quad (6.12)$$

Area of embankment:

$$A_E = \frac{(\tfrac{1}{2}B \pm kH)^2}{2(k \pm s_L)} \quad (6.13a)$$

Area of cutting:

$$A_C = \frac{(\tfrac{1}{2}B \pm kH)^2}{2(k \pm s_R)} \quad (6.13b)$$

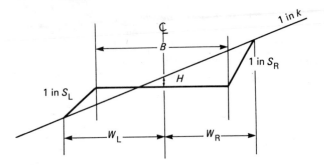

Figure 6.4 Cut and fill cross-section

Example 6.4
Determine the area of the cross-section shown in Figure 6.4 given the following data:

Formation width = 20 m
Formation depth below terrain at centre line = 1.0 m
Side slopes = left − 1 in 2, Right + 1 in 1
Transverse gradients of terrain = Left − 1 in 5, Right + 1 in 5

Solution:
Using Equation (6.12):

$$W_L = \frac{(-5)\times(-2)}{(-5)-(-2)}\left(\frac{20}{2\times(-2)}+(-1)\right) = 13.3\,\text{m}$$

$$W_R = \frac{(+5)\times(+1)}{5-1}\left(\frac{20}{2\times 1}-(-1)\right) = 13.75\,\text{m}$$

Area of embankment:

$$A_E = \frac{((\frac{1}{2}\times 20)-(5\times 1))^2}{2(5-2)} = 4.2\,\text{m}^2$$

Area of cutting:

$$A_C = \frac{((\frac{1}{2}\times 20)+(5\times 1))^2}{2(5-1)} = 28.1\,\text{m}^2$$

Program 6.4 Cross-section area (cut and fill)

```
FORMATION WIDTH 20    FORMATION HEIGHT  -1.00
SIDE SLOPES: LEFT 1 IN -2    RIGHT 1 IN  1
TRANSVERSE GROUND SLOPES LEFT 1 IN -5    RIGHT 1 IN  5

LEFT SIDE WIDTH 13.33  RIGHT SIDE WIDTH 13.75

AREA IN EMBANKMENT  -4.2    AREA IN CUTTING  28.1
```

```
100 PRINT "PROGRAM 6.4"
110 REM THIS PROGRAM CALCULATES THE AREA OF CUT AND FILL IN RESPECT
120 REM OF A SIDE-HILL SECTION (EXAMPLE 6.4)
130 INIT
140 PRINT "ENTER FORMATION WIDTH"
150 INPUT B
160 PRINT "ENTER HORIZONTAL COMPONENT OF GRADIENT OF LEFT SIDE SLOPE"
170 INPUT S1
180 PRINT "ENTER HORIZONTAL COMPONENT OF GRADIENT OF RIGHT SIDE SLOPE"
190 INPUT S2
200 PRINT "ENTER HEIGHT OF FORMATION LEVEL ABOVE(+), OR BELOW(-)"
210 PRINT "GROUND LEVEL AT CENTRE LINE"
220 INPUT H
230 PRINT "ENTER HORIZONTAL COMPONENT OF GRADIENT OF LEFT TRANSVERSE"
240 PRINT "GROUND SLOPE"
250 INPUT K1
260 PRINT "ENTER HORIZONTAL COMPONENT OF GRADIENT OF RIGHT TRANSVERSE";
270 PRINT "GROUND SLOPE"
280 INPUT K2
290 W1=K1*S1/(K1-S1)*(B/(2*S1)-H)
300 W2=K2*S2/(K2-S2)*(B/(2*S2)-H)
310 A1=(0.5*B-K1*H)^2/(2*(K1-S1))
320 A2=(0.5*B-K2*H)^2/(2*(K2-S2))
330 PRINT @4: USING 340:B,H
340 IMAGE 6X,"FORMATION WIDTH",3D,2X,"FORMATION HEIGHT",3D.2D
350 PRINT @4: USING 360:S1,S2
360 IMAGE 6X,"SIDE SLOPES: LEFT 1 IN",3D,3X,"RIGHT 1 IN",3D
370 PRINT @4: USING 380:K1,K2
380 IMAGE 6X,"TRANSVERSE GROUND SLOPES LEFT 1 IN",3D,3X,"RIGHT 1 IN",3D
390 PRINT @4:
400 PRINT @4: USING 410:W1,W2
410 IMAGE 6X,"LEFT SIDE WIDTH",3D.2D,2X,"RIGHT SIDE WIDTH",3D.2D
420 PRINT @4:
430 PRINT @4: USING 440:A1,A2
440 IMAGE 6X,"AREA IN EMBANKMENT",4D.D,5X,"AREA IN CUTTING",4D.D
450 STOP
460 END
```

(c) *Multi-level section*

The most convenient method for determining the area of multi-level sections is to derive coordinates for the points at changes in gradient based, for example, on an origin at the centre of the formation level and using offsets from the centre line as the X coordinates and the relative heights of points above or below the formation level as the Y coordinates (Figure 6.5).

The offsets to the two lateral earthwork extremities (L and R) and their heights may be determined from:

$$W_L = D_{(L-1)} + \left[k_L s \left(H_{(L-1)} - \left(\frac{D_{(L-1)} - B/2}{s} \right) \right) \right] / (s \pm k_L)$$

$$W_R = D_{(R-1)} + \left[k_R s \left(H_{(R-1)} - \left(\frac{D_{(R-1)} - B/2}{s} \right) \right) \right] / (s \pm k_R)$$

(6.14a)

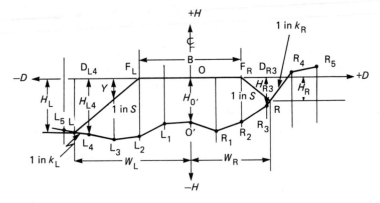

Figure 6.5 Multilevel cross-section

$$H_{\mathrm{L}}=\left(\frac{W_{\mathrm{L}}-B/2}{s}\right)=H_{(\mathrm{L}-1)}\pm\left(\frac{W_{\mathrm{L}}-D_{(\mathrm{L}-1)}}{k_{\mathrm{L}}}\right) \qquad (6.14b)$$

$$H_{\mathrm{R}}=\left(\frac{W_{\mathrm{R}}-B/2}{s}\right)=H_{(\mathrm{R}-1)}\pm\left(\frac{W_{\mathrm{R}}-D_{(\mathrm{R}-1)}}{k_{\mathrm{R}}}\right) \qquad (6.14c)$$

where $D_{(\mathrm{L}-1)}, D_{(\mathrm{R}-1)}$ = the horizontal distances from the centre line to the terrain section points immediately before the points marking the earthwork extremity (points L4 and R3 in Figure 6.5), and $H_{(\mathrm{L}-1)}, H_{(\mathrm{R}-1)}$ = the difference in height between those points and the formation level ($H_{\mathrm{L}4}$ and $H_{\mathrm{R}3}$ in Figure 6.5).

The area of the cross-section may then be determined using coordinates.

Example 6.5
The following data relates to the section shown in Figure 6.5:

Formation width 20 = m
Formation height above terrain at centre line = 6.3 m
Side slopes = Left − 1 in 2, Right − 1 in 2
Heights of terrain section points (5 m intervals from centre line) relative to the formation level

Centre line

	L5	L4	L3	L2	L1	O1	R1	R2	R3	R4	R5
Peg	-6.8	-10.5	-11.3	-10.8	-7.2	-6.3	-6.5	-6.4	-5.8	$+0.8$	$+1.0$

Determine the area of the cross-section.

Solution:

	L5 D_{L5}	L4 D_{L4}	L3 D_{L3}	L2 D_{L2}	L1 D_{L1}	O' O'	R1 D_{R1}	R2 D_{R2}	R3 D_{R3}	R4 D_{R4}	R5 D_{R5}
Peg	-25	-20	-15	-10	-5	0	$+5$	$+10$	$+15$	$+20$	$+25$

Heights

	L5	L4	L3	L2	L1	O1	R1	R2	R3	R4	R5
Formation	$+15$	-10	-05	0	0	0	0	0	-5	-10	-15
Terrain	-6.8	-10.5	-11.3	-10.8	-7.2	-6.3	-6.5	-6.4	-5.8	$+0.8$	$+1.0$

The intersections of the earthwork surface and the terrain surface therefore occur between L4 and L5, and R3 and R4.

$D_{L4} = 20\,\text{m}, \quad H_{L4} = -10.5\,\text{m}$
$D_{L5} = 25\,\text{m}, \quad H_{L5} = -6.8\,\text{m}$
Gradient $L_4 L_5 = 3.7/5 = 1$ in 1.35

$$W_L = D_{L4} + \left(H_{L4} - \frac{10}{2}\right)\left(\frac{k_L \times s}{k_L + s}\right)$$

$$= 20 + \left(10.5 - \frac{10}{2}\right)\left(\frac{1.35 \times 2}{1.35 + 2}\right)$$

$$= 24.43$$

$$H_L = \frac{14.43}{2} = -7.21$$

$D_{R3} = 15\,\text{m} \qquad H_{R3} = -5.8\,\text{m}$
$D_{R4} = 20\,\text{m} \qquad H_{R4} = +0.8\,\text{m}$
Gradient $R_3 R_4 = 6.6/5 = 1$ in 0.76

$$W_R = D_{R3} + \left(H_{R3} - \frac{5}{2}\right)\left(\frac{K_R \times s}{K_R + s}\right)$$

$$= 15 + \left(5.8 - \frac{5}{2}\right)\left(\frac{0.76 \times 2}{0.76 + 2}\right)$$

$$= 16.82$$

$$H_R = \frac{6.82}{2} = -3.41$$

Area by coordinates:

	D		H
		$(+24.43)$	
O	0	$+24.43$	0
F_L	-10	$+14.43$	0
L	-24.43	0	-7.21
L_4	-20	$+4.43$	-10.5
L_3	-15	$+9.43$	-11.3
L_2	-10	$+14.43$	-10.8
L_1	-5	$+19.43$	-7.2
O'	0	$+24.43$	-6.3
R_1	$+5$	$+29.43$	-6.5
R_2	$+10$	$+34.43$	-6.4
R_3	$+15$	$+39.43$	-5.8
R	$+16.82$	$+41.25$	-3.41
F_R	$+10$	$+34.43$	0

Area $= 534.4/2 = 267.2\,\text{m}^2$
Check $534.4/2 = 267.2\,\text{m}^2$

Program 6.5 Cross-section area (multi-level solution)

H	-6.8	-10.5	-11.3	-10.8	-7.2	-6.3	-6.5	-6.4	-5.8	0.8	1.0
X	-25.0	-20.0	-15.0	-10.0	-5.0	0	5.0	10.0	15.0	20.0	25.0
Y/N	-7.5	-5.0	-2.5	0.0	0.0	0	0.0	0.0	-2.5	-5.0	-7.5

```
COORDINATES OF TURNING POINTS

              X                          Y

LEFT SIDE

    1        -5.0                      -7.2
    2       -10.0                     -10.8
    3       -15.0                     -11.3
    4       -20.0                     -10.5
    5       -24.4                      -7.2
    6       -10.0                       0.0
    7         0.0                       0.0
    8         0.0                      -6.3

RIGHT SIDE

    1         5.0                      -6.5
    2        10.0                      -6.4
    3        15.0                      -5.8
    4        16.8                      -3.4
    5        10.0                       0.0
    6         0.0                       0.0
    7         0.0                      -6.3

AREA LEFT SIDE       175.70

AREA RIGHT SIDE       91.49

TOTAL AREA           267.19


100 PRINT "PROGRAM 6.5"
110 REM THIS PROGRAM DETERMINES THE AREA OF A MULTI LEVEL ROAD EARTHWORK
120 REM CROSS SECTION BY MEANS OF COORDINATES (EXAMPLE 6.5)
130 INIT
140 PRINT "ENTER FORMATION WIDTH"
150 INPUT B
160 PRINT "ENTER HEIGHT OF FORMATION ABOVE TERRAIN ON CENTRE LINE"
170 INPUT H
180 PRINT "ENTER HORIZONTAL COMPONENT OF GRADIENT OF LEFT SIDE SLOPE"
190 INPUT S1
200 PRINT "ENTER HORIZONTAL COMPONENT OF GRADIENT OF RIGHT SIDE SLOPE"
210 INPUT S2
220 PRINT "ENTER SPACING OF SECTION POINTS"
230 INPUT D
240 PRINT "ENTER NUMBER OF SECTION POINTS LEFT OF CENTRE LINE"
250 INPUT P
260 PRINT "ENTER NUMBER OF SECTION POINTS RIGHT OF CENTRE LINE"
270 INPUT Q
280 DIM L(P+3),R(Q+3),X(P+3),E(Q+3),Y(P),N(Q)
290 X(1)=-D
300 PRINT "ENTER HEIGHT OF TERRAIN SECTION POINT LEFT 1"
310 INPUT L(1)
320 E(1)=D
330 PRINT "ENTER HEIGHT OF TERRAIN SECTION POINT RIGHT 1"
340 INPUT R(1)
350 FOR I=2 TO P
360 PRINT "ENTER HEIGHT OF TERRAIN SECTION POINT LEFT ";I
370 INPUT L(I)
380 X(I)=X(I-1)-D
390 NEXT I
400 FOR J=2 TO Q
410 PRINT "ENTER HEIGHT OF TERRAIN SECTION POINT RIGHT ";J
```

```
420 INPUT R(J)
430 E(J)=E(J-1)+D
440 NEXT J
450 PRI 94: USI 460:L(5),L(4),L(3),L(2),L(1),-H,R(1),R(2),R(3),R(4),R(5)
460 IMAGE 2X,"H",5X,11(3D.D,X)
470 PRINT 94:
480 PRI 94: USI 490:X(5),X(4),X(3),X(2),X(1),E(1),E(2),E(3),E(4),E(5)
490 IMAGE  2X,"X",5X,5(3D.D,X),2X,"0",3X,5(3D.D,X)
500 PRINT 94:
510 Y(1)=0
520 N(1)=0
530 Y(2)=0
540 N(2)=0
550 FOR K=3 TO P
560 Y(K)=Y(K-1)+D/S1
570 N(K)=N(K-1)+D/S2
580 NEXT K
590 PRI 94: USI 600:Y(5),Y(4),Y(3),Y(2),Y(1),N(1),N(2),N(3),N(4),N(5)
600 IMAGE 2X,"Y/N",3X,5(3D.D,X),2X,"0",3X,5(3D.D,X)
610 FOR V=1 TO P
620 U1=Y(V)-L(V)
630 IF U1<1 THEN 690
640 NEXT V
650 FOR W=1 TO Q
660 U2=N(W)-R(W)
670 IF U2<1 THEN 810
680 NEXT W
690 F1=ABS((ABS(X(V))-ABS(X(V-1)))/(ABS(L(V))-ABS(L(V-1))))
700 G1=ABS(X(V-1))
710 G1=G1+(ABS(L(V-1))-ABS(Y(V-1)))*(F1*ABS(S1))/(F1+ABS(S1))
720 X(V)=G1*SGN(X(V-1))
730 L(V)=(G1-ABS(X(2)))/ABS(S1)*SGN(L(V-1))
740 X(V+1)=X(2)
750 X(V+2)=0
760 X(V+3)=0
770 L(V+1)=0
780 L(V+2)=0
790 L(V+3)=-H
800 GO TO 650
810 F2=ABS((ABS(E(W))-ABS(E(W-1)))/(ABS(R(W))+ABS(R(W-1))))
820 G2=ABS(E(W-1))
830 G2=G2+(ABS(R(W-1))-ABS(N(W-1)))*(F2*ABS(S2))/(F2+ABS(S2))
840 E(W)=G2*SGN(E(W-1))
850 R(W)=(G2-ABS(E(2)))/ABS(S2)*SGN(R(W-1))
860 E(W+1)=E(2)
870 E(W+2)=0
880 E(W+3)=0
890 R(W+1)=0
900 R(W+2)=0
910 R(W+3)=-H
920 PRINT 94:
930 PRINT 94:
940 PRINT 94:"COORDINATES OF TURNING POINTS"
950 PRINT 94:
960 PRINT 94: USING 970:
970 IMAGE 11X,"X",25X,"Y"
980 PRINT 94:
990 PRINT 94:"LEFT SIDE"
1000 PRINT 94:
1010 FOR I=1 TO V+3
1020 PRINT 94: USING 1030:I,X(I),L(I)
1030 IMAGE 2X,D,5X,4D.D,20X,4D.D
1040 NEXT I
1050 PRINT 94:
1060 PRINT 94:"RIGHT SIDE"
1070 PRINT 94:
1080 FOR I=1 TO W+3
1090 PRINT 94: USING 1030:I,E(I),R(I)
1100 NEXT I
1110 FOR I=1 TO V+3
1120 X(I)=ABS(X(I))
1130 L(I)=ABS(L(I))
1140 NEXT I
```

```
1150 A1=0
1160 J1=X(2)-X(V+3)
1170 K1=J1*L(1)
1180 A1=A1+K1
1190 FOR I=2 TO V+2
1200 J1=X(I+1)-X(I-1)
1210 K1=J1*L(I)
1220 A1=A1+K1
1230 NEXT I
1240 J1=X(1)-X(V+2)
1250 K1=J1*L(V+3)
1260 A1=A1+K1
1270 PRINT #4:
1280 PRINT #4: USING 1290:A1/2
1290 IMAGE "AREA LEFT SIDE",10D.2D
1300 FOR I=1 TO W+3
1310 E(I)=ABS(E(I))
1320 R(I)=ABS(R(I))
1330 NEXT I
1340 A2=0
1350 J2=E(2)-E(W+3)
1360 K2=J2*R(1)
1370 A2=A2+K2
1380 FOR I=2 TO W+2
1390 J2=E(I+1)-E(I-1)
1400 K2=J2*R(I)
1410 A2=A2+K2
1420 NEXT I
1430 J2=E(1)-E(W+2)
1440 K2=J2*R(W+3)
1450 A2=A2+K2
1460 PRINT #4:
1470 PRINT #4: USING 1480:A2/2
1480 IMAGE "AREA RIGHT SIDE",9D.2D
1490 A3=A1/2+A2/2
1500 PRINT #4:
1510 PRINT #4: USING 1520:A3
1520 IMAGE "TOTAL AREA",14D.2D
1530 STOP
1540 END
```

6.1.2 Volume determination

The determination of volume is generally approached by breaking down the total volume of the solid into blocks that are contained between parallel planes. Artificial surfaces in engineering works are frequently plane and so the accuracy of, for instance, earthwork volumes, depends on the density of spot heights being properly related to the relief of the ground, so that surfaces between adjacent points of known height may be considered plane.

6.1.2.1 *Volumes from cross-sections*

Volumes between vertical cross-sections arise in earthwork calculations for road and rail alignments. The cross-sections are taken at right angles to the centreline of the alignment and are therefore parallel to one another except on curves. Where adjacent

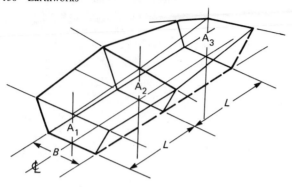

Figure 6.6 The prismoid in highway earthworks

cross-sections of a straight cutting or embankment are parallel the solid of excavation or fill between them approximates to the geometrical form of a prismoid having end faces in parallel planes. These faces are polygons not necessarily with the same number of sides. The longitudinal or side faces are all plane, consisting of triangles or trepezia formed by straight lines joining corners on opposite end faces (Figure 6.6).

Assuming that cross-sections are all parallel and equally spaced, the volume (V) of the earthwork solid may be determined in several ways:

(*a*) *End areas:*

$$V = L\left[\frac{1}{2}(A_1 + A_n) + \sum_{(n-1)}^{i=2} A_i\right] \qquad (6.15)$$

where L = the distance between adjacent cross-sections, A = the area of cross-sections, and subscripts = cross section number up to n.

(*b*) *Prismoidal formula*
Assuming that the earth solid between cross-sections approximates in form to a prismoid its volume is given by

$$V = \frac{L}{6}(A_1 + 4A_m + A_2) \qquad (6.16)$$

where A_m is the area of the mid-section and *not* the average of the end areas unless the prismoid is composed solely of prisms and wedges (the mid-section area of a pyramid $= \frac{1}{4} \times$ basal area and so in general the mid-section area of a prismoid is not the mean of the end areas). The problem is how to deal with A_m.

One solution is to add cross-sections mid-way between the regular sections which are L apart but this would double the field work in levelling and pegging cross-sections.

An alternative is to treat every other regular section A as mid-section, doubling the length of each prismoid. Hence

$$V = \frac{L}{3}[A_1 + 4(A_2 + A_4 + A_6 \ldots A_{n-1})$$
$$+ 2(A_3 + A_5 + A_7 + \ldots A_{n-2}) + A_n] \qquad (6.17)$$

Example 6.6
Referring to Figure 6.6 let the areas of cross-sections spaced 50 m apart be as follows:

$A_1 = 238 \, m^2$
$A_2 = 387 \, m^2$
$A_3 = 302 \, m^2$

Determine the volume of the cutting.

Solution:
Using Equation (6.15):

$$V = 50[\tfrac{1}{2}(238 + 302) + 387]$$
$$= 32\,850 \, m^3$$

Using Equation (6.16) and assuming $A_2 = A_m$:

$$V = \frac{100}{6}(238 + (4 \times 387) + 302)$$
$$= 34\,800 \, m^3$$

Program 6.6 Volumes by end areas/prismoidal formulae

```
AREA OF SECTION 1 = 238
AREA OF SECTION 2 = 387
AREA OF SECTION 3 = 302


VOLUME (END AREA FORMULA)       32850
VOLUME (PRISMOIDAL FORMULA)     34800
```

```
100 PRINT "PROGRAM 6.6"
110 REM THIS PROGRAM DETERMINES THE VOLUME OF THE EARTH SOLID REFERRED
120 REM TO IN EXAMPLE 6.6 USING THE END AREAS AND PRISMOIDAL FORMULAE
130 REM (ASSUMING THE CENTRAL SECTION TO EQUAL THE MEAN OF THE TWO END
140 REM SECTIONS)
150 INIT
160 DIM A(3)
170 READ A(1),A(2),A(3),X
180 DATA 238,387,302,50
190 V1=X*(0.5*(A(1)+A(3))+A(2))
200 V2=2*X/6*(A(1)+4*A(2)+A(3))
210 FOR I=1 TO 3
220 PRINT @4:"AREA OF SECTION ";I;" = ";A(I)
230 NEXT I
240 PRINT @4:
250 PRINT @4: USING 260:V1
260 IMAGE"VOLUME (END AREA FORMULA)",2X,7D
270 PRINT @4: USING 280:V2
280 IMAGE"VOLUME (PRISMOIDAL FORMULA)",7D
290 STOP
300 END
```

A third solution is to interpolate mid-section linear dimensions as the mean of corresponding end section dimensions. In practice this is carried out indirectly in two stages, first computing a value V' for the volume by the end areas formulae and then applying a 'prismoidal correction' to V' to determine V.

Prismoidal correction
In general $V' > V$. Therefore V' requires a negative prismoidal correction (PC). In respect of the earth solid between two cross-sections:

$$PC = V' - V = \frac{L}{12}(H_2 - H_1)[(W_{R2} + W_{L2}) - (W_{R1} + W_{L1})] \quad (6.18)$$

where H = formation depths at centreline, W_R = right side width, W_L = left side width, and subscripts 1 and 2 = cross-section numbers.

Example 6.7
Referring to Figures 6.6 and 6.7, determine the prismoidal corrections for the two sections and hence the volume of the cutting.

Solution:

Section one:

$$V_1' = \left(\frac{238 + 387}{2}\right) \times 50 = 15\,625\,\text{m}^3$$

$$PC = \frac{50}{12}(7.8 - 7.0)[(17.1 + 51.2) - (24.0 + 24.0)]$$
$$= 67\,\text{m}^3$$
$$V_1 = 15\,625 - 67 = 15\,558\,\text{m}^3$$

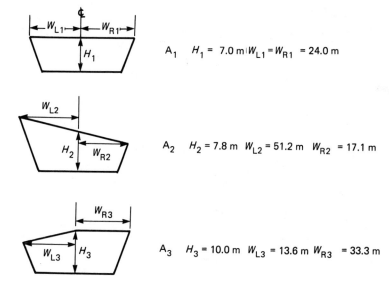

Figure 6.7 Sidewidths in highway cross-sections

Section two:

$$V_2' = \left(\frac{387+302}{2}\right) \times 50 = 17\,225 \text{ m}^3$$

$$\text{PC} = \frac{50}{12}(10.0-7.8)[(33.3+13.6)-(17.1+51.2)]$$
$$= -196.2$$
$$V_2 = 17\,225 + 196.1 = 17\,421 \text{ m}^3$$
Total volume $= V_1 + V_2 = 32\,979 \text{ m}^3$.

Using the prismoidal formula for each section:

Section one:

$$H_m = 7.4, \ W_{L_m} = 37.6, \ W_{R_m} = 20.55$$
$$A_m = (\tfrac{1}{2} \times 58.15)\left(7.4 + \frac{20}{2 \times 2}\right) - \frac{20^2}{4 \times 2} = 310.5$$
$$V_1 = \frac{50}{6}(238 + (4 \times 310.5) + 387) = 15\,558 \text{ m}^3 \quad \text{(check)}$$

Section two:

$$H_m = 8.9, \quad W_{L_m} = 32.4, \quad W_{R_m} = 25.2$$

$$A_m = (\tfrac{1}{2} \times 57.6)\left(8.9 + \frac{20}{2 \times 2}\right) - \frac{20^2}{4 \times 2} = 350.3$$

$$V_2 = \frac{50}{6}(387 + (4 \times 350.3) + 302) = 17\,418\,\text{m}^3 \quad \text{(check)}$$

Program 6.7 Volume using prismoidal correction

```
SECTION 1

VOLUME (END AREAS)          15625
PRISMOIDAL CORRECTION         -68
CORRECTED VOLUME            15557

SECTION 2

VOLUME (END AREAS)          17225
PRISMOIDAL CORRECTION         196
CORRECTED VOLUME            17421


TOTAL VOLUME                32979


100 PRINT "PROGRAM 6.7"
110 REM THIS PROGRAM DETERMINES THE VOLUME OF THE EARTH SOLID REFERRED
120 REM TO IN EXAMPLE 6.6 BY APPLICATION OF THE PRISMOIDAL CORRECTION
130 REM TO THE END AREAS VOLUME (EXAMPLE 6.7)
140 INIT
150 DIM A(3),H(3),L(3),R(3),V(2),C(2)
160 READ A(1),A(2),A(3),X,H(1),H(2),H(3),L(1),L(2),L(3),R(1),R(2),R(3)
170 DATA 238,387,302,50,7,7.8,10,24,51.2,13.6,24,17.1,33.3
180 T=0
190 FOR I=1 TO 2
200 PRINT @4:
210 PRINT @4:"SECTION ";I
220 PRINT @4:
230 V(I)=(A(I)+A(I+1))/2*X
240 PRINT @4: USING 250:V(I)
250 IMAGE"VOLUME (END AREAS)",6X,7D
260 C(I)=-X/12*(H(I+1)-H(I))*(R(I+1)+L(I+1)-(R(I)+L(I)))
270 PRINT @4: USING 280:C(I)
280 IMAGE "PRISMOIDAL CORRECTION",4X,6D
290 V(I)=V(I)+C(I)
300 PRINT @4: USING 310:V(I)
310 IMAGE "CORRECTED VOLUME",15D
320 T=T+V(I)
330 NEXT I
340 PRINT @4:
350 PRINT @4:
360 PRINT @4: USING 370:T
370 IMAGE "TOTAL VOLUME",19D
380 STOP
390 END
```

For side-hill sections:

$$\text{PC}_{(CUT)} = -\frac{L}{12}(H_{R1} - H_{R2})(D_1 - D_2) \tag{6.19a}$$

$$PC_{(FILL)} = \frac{L}{12}(H_{L1} - H_{L2})(D_1 - D_2) \qquad (6.19b)$$

where H_R = difference in height between the formation level and the right wing point, H_L = difference in height between the formation level and the left wing point, D = distance from the centreline to the point of intersection between the formation level and the existing transverse slope of the terrain. If the centreline lies in fill, the signs of D_1 and D_2 must be reversed.

Example 6.8
Using the following data determine the prismoidal correction for the section of earthworks illustrated in Figure 6.8.

Distance between cross-sections = 50 m
Formation width = 20 m
Formation depth below terrain at centre line =
  Section 1   1.0 m   Section 2   2.2 m.
Side slopes = Left − 1 in 2   Right + 1 in 1
Transverse gradients of terrain =
  Section 1   Left − 1 in 5   Right + 1 in 5
  Section 2   Left − 1 in 4   Right + 1 in 4

Solution:

$D_1 = 1.0 \times 5 = 5.0$ m
$D_2 = 2.2 \times 4 = 8.8$ m
$W_{L1} = 13.3$ m $W_{R1} = 13.75$ m (from Example 6.4)

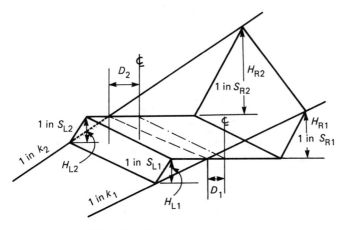

Figure 6.8 Cut and fill earth solid

$$W_{L2} = \left(\frac{-4 \times -2}{4-2}\right)\left(\frac{20}{2 \times 2} - 1\right), \quad W_{R2} = \left(\frac{+4 \times +1}{4-1}\right)\left(\frac{20}{2 \times 1} + 1\right)$$

$$= 16.0 \, \text{m} \qquad\qquad\qquad = 14.7 \, \text{m}$$

$$H_{R1} = (13.75 + 5.0)/5 = 3.75$$
$$H_{L1} = (13.3 - 5.0)/5 = 1.66$$
$$H_{R2} = (14.7 + 8.8)/4 = 5.88$$
$$H_{L2} = (16.0 - 8.8)/4 = 1.80$$
$$PC_{(CUT)} = -\frac{50}{12}(3.75 - 5.88)(5.0 - 8.8)$$
$$= -33.7 \, \text{m}^3$$
$$PC_{(FILL)} = \frac{50}{12}(1.66 - 1.80)(5.0 - 8.8)$$
$$= 2.2 \, \text{m}^3$$

Program 6.8 Volume using prismoidal correction (cut and fill)

```
SPACING OF CROSS SECTIONS 50
FORMATION WIDTH 20
FORMATION DEPTH BELOW TERRAIN AT CENTRE LINE SECTION 1 1.0; SECTION 2 2.2
TERRAIN TRANSVERSE SLOPE SECTION 1  LEFT -1:5 ; RIGHT +1:5
                         SECTION 2  LEFT -1:4 ; RIGHT +1:4
SIDE SLOPES                         LEFT -1:2 ; RIGHT -1:2
SIDEWIDTHS SECTION 1 LEFT 1.66; RIGHT 3.75
           SECTION 2 LEFT 1.80; RIGHT 5.88


PRISMOIDAL CORRECTIONS CUT 34  FILL  2


100 PRINT "PROGRAM 6.8"
110 REM THIS PROGRAM DETERMINES THE PRISMOIDAL CORRECTIONS APPLICABLE
120 REM TO THE EARTH SOLID REFERRED TO IN EXAMPLE 6.8
130 INIT
140 DIM D(2),K(2),L(2),R(2),N(2),M(2),C(2),H(2)
150 READ X,B,H(1),H(2),L(1),L(2),R(1),R(2),K(1),K(2)
160 DATA 50,20,1,2.2,13.3,16,13.75,14.7,5,4
170 FOR I=1 TO 2
180 D(I)=H(I)*K(I)
190 N(I)=(L(I)-D(I))/K(I)
200 M(I)=(R(I)+D(I))/K(I)
210 NEXT I
220 C(1)=X/12*((M(1)-M(2))*(D(1)-D(2)))
230 C(2)=X/12*((N(1)-N(2))*(D(1)-D(2)))
240 PRINT @4:"SPACING OF CROSS SECTIONS ";X
250 PRINT @4:"FORMATION WIDTH ";B
260 PRINT @4:"FORMATION DEPTH BELOW TERRAIN AT CENTRE LINE SECTION 1";
270 PRINT @4:" 1.0; SECTION 2 2.2"
280 PRI @4:"TERRAIN TRANSVERSE SLOPE SECTION 1  LEFT -1:5 ; RIGHT +1:5"
290 PRI @4:"                         SECTION 2  LEFT -1:4 ; RIGHT +1:4"
300 PRI @4:"SIDE SLOPES                         LEFT -1:2 ; RIGHT -1:2"
310 PRINT @4:"SIDEWIDTHS SECTION 1 LEFT 1.66; RIGHT 3.75"
320 PRINT @4:"           SECTION 2 LEFT 1.80; RIGHT 5.88"
330 PRINT @4:
340 PRINT @4:
350 PRINT @4: USING 360:C(1),C(2)
360 IMAGE "PRISMOIDAL CORRECTIONS CUT ",2D,2X,"FILL ",2D
370 STOP
380 END
```

Curvature corrections
Where the centre line is not straight but lies on a curve of radius R, the volume as calculated using the end areas of prismoidal formula must be corrected as the cross-sections are no longer parallel (Figure 6.9(a)).

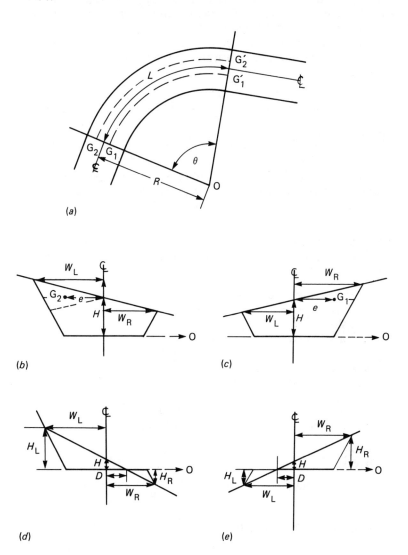

Figure 6.9 The curvature correction in highway earthwork volume determination

The magnitude of the correction may be determined from:

$$C = \pm \frac{L}{6R}(W_R^2 - W_L^2)\left(H + \frac{B}{2S}\right) \tag{6.20}$$

The sign of the correction is positive/negative if the centroid of the cross-section, e, lies further from/closer to the centre of the circular arc than the centre line (Figures 6.9(b) and 6.9(c)).

For side-hill sections with the centre line in cut and the transverse gradient sloping downwards towards the centre of the curve (Figure 6.9(d)):

$$C_{(\text{CUT})} = \frac{L}{3R}\left(W_L + \frac{B}{2} - D\right)\left[\frac{H_L}{2}\left(\frac{B}{2} + D\right) - HD\right] \tag{6.21a}$$

$$C_{(\text{FILL})} = \frac{L}{3R}\left(W_R + \frac{B}{2} + D\right)\left[\frac{H_R}{2}\left(\frac{B}{2} - D\right)\right] \tag{6.21b}$$

For the centre line in fill the two formulae are interchanged and if the transverse gradient slopes away from the centre of the curve, the signs are reversed (Figure 6.9(e)).

Example 6.9
Given the following data determine the volume of the cutting shown in Figure 6.9(a) assuming that the end sections are of the form shown in Figure 6.9(b):

Interval between sections	50 m	
Formation width	20 m	
Side slopes	1 in 2	
Radius of curve	800 m	

	Section 1	Section 2
Transverse slope of terrain	1 in 10	1 in 5
Depth of centre line below terrain	5 m	10 m
Side widths	R 16.67	R 21.43
	L 25.00	L 50.00

Solution:
   Areas:

Section 1 $\frac{1}{2}(16.67 + 25.00)\left(5 + \frac{20}{4}\right) - \frac{400}{8} = 158.34\,\text{m}^2$

Section 2 $\frac{1}{2}(21.43 + 50.00)\left(10 + \frac{20}{4}\right) - \frac{400}{8} = 485.72\,\text{m}^2$

Volume (end areas) $= \frac{50}{2}(158.34 + 485.72) = 16\,101.50\,\text{m}^3$

Correction:

$$\text{Section 1} \quad \frac{50}{12 \times 800}(25^2 - 16.67^2)\left(5 + \frac{20}{2 \times 2}\right) = 18.08 \text{ m}^3$$

$$\text{Section 2} \quad \frac{50}{12 \times 800}(50^2 - 21.43^2)\left(10 + \frac{20}{2 \times 2}\right) = 159.43 \text{ m}^3$$

Total volume correction = 177.51 m^3
Corrected volume = 16 101.50 + 177.51 = 16 279 m^3

Program 6.9 *Volume using curvature correction*

```
SECTION INTERVAL 50
FORMATION WIDTH 20
SIDE SLOPES  1 IN 2
RADIUS OF CURVE 800


SECTION 1

TRANSVERSE SLOPE OF TERRAIN 1 IN 10
DEPTH OF CENTRE LINE 5
SIDE WIDTHS: LEFT 25 RIGHT 16.67

SECTION 2

TRANSVERSE SLOPE OF TERRAIN 1 IN 5
DEPTH OF CENTRE LINE 10
SIDE WIDTHS: LEFT 50 RIGHT 21.43


AREA SECTION 1 158.35
AREA SECTION 2 485.725

CURVATURE CORRECTION TO VOLUME SECTION 1      18
CURVATURE CORRECTION TO VOLUME SECTION 2     159

VOLUME (END AREAS)   16102
VOLUME CORRECTED FOR CURVATURE    16279

100 PRINT "PROGRAM 6.9"
110 REM THIS PROGRAM DETERMINES THE VOLUME OF AN EARTH SOLID BY APPLYING
120 REM A CURVATURE CORRECTION TO THE END AREAS VOLUME(EXAMPLE 6.9)
130 INIT
140 DIM A(2),R(2),L(2),D(2),C(2),K(2)
150 READ X,Y,B,S,K(1),K(2),L(1),L(2),R(1),R(2),D(1),D(2)
160 DATA 50,800,20,2,10,5,25,50,16.67,21.43,5,10
170 V=0
180 FOR I=1 TO 2
190 A(I)=0.5*(R(I)+L(I))*(D(I)+B/(2*S))-B^2/(4*2)
200 C(I)=X/(12*Y)*(L(I)^2-R(I)^2)*(D(I)+B/(2*S))
210 V=V+C(I)
220 NEXT I
230 V=X/2*(A(1)+A(2))+V
240 PRINT @4:"SECTION INTERVAL ";X
250 PRINT @4:"FORMATION WIDTH ";B
260 PRINT @4:"SIDE SLOPES  1 IN ";S
270 PRINT @4:"RADIUS OF CURVE ";Y
280 PRINT @4:
290 PRINT @4:
300 PRINT @4:"SECTION 1 "
310 PRINT @4:
320 PRINT @4:"TRANSVERSE SLOPE OF TERRAIN 1 IN ";K(1)
330 PRINT @4:"DEPTH OF CENTRE LINE ";D(1)
340 PRINT @4:"SIDE WIDTHS: LEFT ";L(1);" RIGHT ";R(1)
350 PRINT @4:
```

```
360 PRINT @4:"SECTION 2"
370 PRINT @4:
380 PRINT @4:"TRANSVERSE SLOPE OF TERRAIN 1 IN ";K(2)
390 PRINT @4:"DEPTH OF CENTRE LINE ";D(2)
400 PRINT @4:"SIDE WIDTHS: LEFT ";L(2);" RIGHT ";R(2)
410 PRINT @4:
420 PRINT @4:
430 PRINT @4:"AREA SECTION 1 ";A(1)
440 PRINT @4:"AREA SECTION 2 ";A(2)
450 PRINT @4:
460 PRINT @4: USING 470:C(1)
470 IMAGE "CURVATURE CORRECTION TO VOLUME SECTION 1 ",7D
480 PRINT @4: USING 490:C(2)
490 IMAGE "CURVATURE CORRECTION TO VOLUME SECTION 2 ",7D
500 PRINT @4:
510 PRINT @4: USING 520:X/2*(A(1)+A(2))
520 IMAGE "VOLUME (END AREAS) ",7D
530 PRINT @4: USING 540:V
540 IMAGE "VOLUME CORRECTED FOR CURVATURE ",7D
550 STOP
560 END
```

6.1.2.2 *Volumes from spot heights*

Where, for example, the earthwork is to be an extensive excavation, the surface of the site may be divided into a number of equal triangles by observing a series of spot levels in the form of a grid pattern (Figure 6.10). The density of spot levels is such that the ground surface in each triangle may be regarded as a sloping plane. Then the volume between the ground plane of each triangle and the formation level below it is that of a truncated triangular prism with non-parallel plane ends. If the area of the square or right section of the prism is A

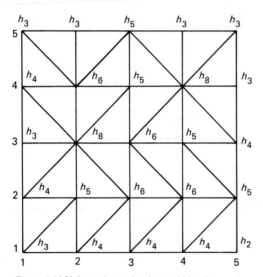

Figure 6.10 Volume determination: grid levels

and the lengths of the three vertical edges are h_1, h_2 and h_3 (Figure 6.11(*a*)), the volume of a single truncated triangular prism is given exactly by:

$$V = \frac{A}{3}(h_1 + h_2 + h_3)$$

This formula is equally applicable to the case where the formation surface is not level (Figure 6.11(*b*)).

The formula may be further extended to cater for more extensive areas which may be divided as illustrated in Figure 6.6, that is:

$$V = \frac{A}{3}\left(\sum_{i=1}^{8} ih_i \right)$$

where h_2 = the total sum of all heights used twice, h_3 = the total sum of all heights used three times, etc.

Example 6.10
The area bounded by the points 1, 1; 5, 1; 5, 5; 1, 5 in Figure 6.10 is to be excavated to provide a level surface at a depth of 10 m below point 1, 1.

Using the following data extracted from the grid of spot levels determine the volume of excavation required in the area defined by points 1, 1; 1, 3; 3, 3; 3, 1.

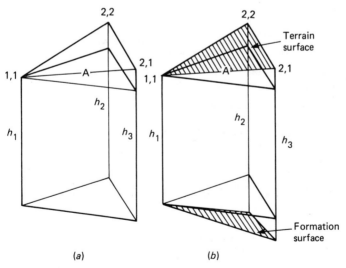

Figure 6.11 Vertical prismoid

Grid interval 5 m

	1,1	1,2	1,3	2,1	2,2	2,3	3,1	3,2	3,3
Altitude (m)	116.2	122.3	125.1	120.0	118.6	122.2	123.6	123.8	125.6

Solution:

$$\text{Area of triangle} = 5 \times \frac{5}{2} = 12.5 \, \text{m}^2$$

Required altitude of finished surface = $116.2 - 10.0 = 106.2$

h	10.0	16.1	18.9	13.8	12.4	16.0	17.4	17.6	19.4
i	2	3	1	3	5	4	1	4	1
ih	20.0	48.3	18.9	41.4	62.0	64.0	17.4	70.4	19.4

$\Sigma ih = 361.8$

$V = A/3 \times \Sigma ih = 12.5/3 \times 361.8 = 1507.5 \, \text{m}^3$

Program 6.10 *Volume from spot heights*

```
GRID INTERVAL 5
TRIANGLE AREA 12.5

POINT          H           I            IH

1,1          10.0         2            20
1,2          16.1         3            48
1,3          18.9         1            19
2,1          13.8         3            41
2,2          12.4         5            62
2,3          16.0         4            64
3,1          17.4         1            17
3,2          17.6         4            70
3,3          19.4         1            19

TOTAL VOLUME 1507.5
```

```
100 PRINT "PROGRAM 6.10"
110 REM THIS PROGRAM DETERMINES THE VOLUME OF AN EARTH SOLID GIVEN
120 REM GRIDDED SPOT HEIGHTS (EXAMPLE 6.10)
130 INIT
140 DIM H(3,3),P(3,3),Q(3,3),N(3,3),S(3,3)
150 G=5
160 PRINT @4:"GRID INTERVAL ";G
170 A=G*G/2
180 PRINT @4:"TRIANGLE AREA ";A
190 PRINT @4:
200 PRINT @4: USING 210:
210 IMAGE X,"POINT",8X,"H",9X,"I",10X,"IH"
220 PRINT @4:
230 T=0
240 FOR I=1 TO 3
250 FOR J=1 TO 3
260 READ P(I,J)
270 DATA 116.2,122.3,125.1,120,118.6,122.2,123.6,123.8,125.6
280 PRINT "ENTER NUMBER OF TRIANGLES CONVERGING ON STATION";I;",";J
290 INPUT N(I,J)
300 Z=P(1,1)-10
310 H(I,J)=P(I,J)-Z
320 Q(I,J)=H(I,J)*N(I,J)
330 PRINT @4: USING 340:I,J,H(I,J),N(I,J),Q(I,J)
340 IMAGE 2X,D,",",D,7X,3D.D,7X,D,7X,5D
350 T=T+Q(I,J)
360 NEXT J
370 NEXT I
```

```
380 V=A/3*T
390 PRINT @4:
400 PRINT @4:"TOTAL VOLUME ";V
410 STOP
420 END
```

6.1.2.3 Volumes from contour lines

Volumes may be determined by using the contour lines of the solid to be measured. The assumption is made that the surface of the solid slopes at a uniform gradient from one contour to the next. Consequently, the smaller the contour interval the more valid is the assumption.

The volume is determined by regarding the solid as being divided into a number of horizontal slices by the contour planes. The depths of these are known, and their end areas (i.e. the areas contained within each contour) are obtained usually by planimeter.

The total volume may therefore be calculated by the prismoidal formula or the end area formula (see Section 6.1.2.3). In using the former either every second area is treated as a mid-area or the mid-areas must be determined from contour lines interpolated midway between each original pair of contours.

6.1.2.4 Mass-haul volumes

The cost of earthworks in highway and railway construction depends largely on the amount and nature of the material involved in the earthworks and the distance over which such material must be moved. The mass-haul diagram is used to determine the limits of excavation and embankment and the direction in which material should be moved, the aim being to minimize the amount of haul (the product of the volume of excavated material by the distance moved to the location of fill) and to obtain a balance of volumes of cut and fill. Should the haul distance be excessive it may be more economical to dump ('waste') material at one location and 'borrow' material at another location.

In order to compile such a diagram cumulative volumes are first determined. When evaluating the cumulative volumes cut quantities are regarded as positive (indicating a surplus of material) and fill as negative (indicating a deficiency of material). The quantities of fill are normally adjusted to allow for the change in volume which usually occurs following excavation of material ('bulking' or 'shrinkage') and the adjusted values of fill are used in determining the cumulative volumes.

The data for the mass haul curve is therefore derived simply from

$$V_A = \sum_{i=1}^{n} V_i$$

where V_A = the cumulative volume at the end of n sections in the alignment, V_i = the volume of an individual section adjusted where necessary for bulking/shrinkage.

Example 6.11
The following figures show the cut ($+$) and fill ($-$) in cubic metres between successive stations 50 m apart in a proposed road alignment.

0	1	2	3	4	5	6	7	8
$+1500$	$+1100$	$+500$	$+100$	-100	-1000	-2200	-2500	

8	9	10	11	12
-1600	-400	$+1800$	$+2800$	

Assuming a correction factor of 0.9 to be applied to fill material for bulking/shrinkage, determine accumulated volumes for use in the preparation of a mass-haul curve.

Solution:

Volume

Station	Cut (m³)	Fill (m³) uncorrected	corrected	Accumulated volume (m³)
0				
	$+1500$			$+1500$
1				
	$+1100$			$+2600$
2				
	$+500$			$+3100$
3				
	$+100$			$+3200$
4				
		-100	-90	$+3110$
5				
		-1000	-900	$+2210$
6				
		-2200	-1980	$+230$
7				
		-2500	-2250	-2020
8				
		-1600	-1440	-3460
9				
		-400	-360	-3820
10				
	$+1800$			-2020
11				
	$+2800$			$+780$
12				

Program 6.11 Accumulation of volumes (mass haul)

SECTION		CUT	FILL		V
0	1	1500			1500
1	2	1100			2600
2	3	500			3100
3	4	100			3200
4	5		-100	-90	3110
5	6		-1000	-900	2210
6	7		-2200	-1980	230
7	8		-2500	-2250	-2020
8	9		-1600	-1440	-3460
9	10		-400	-360	-3820
10	11	1800			-2020
11	12	2800			780

```
100 PRINT "PROGRAM 6.11"
110 REM THIS PROGRAM DETERMINES ACCUMULATED VOLUME REQUIRED FOR THE
120 REM PREPARATION OF A MASS-HAUL CURVE (EXAMPLE 6.11)
130 INIT
140 N=12
150 V=0
160 DIM C(N),F(N)
170 PRINT @4: USING 180:
180 IMAGE 4X,"SECTION",6X,"CUT",11X,"FILL",11X,"V"
190 PRINT @4:
200 FOR I=1 TO N
210 READ C(I)
220 DATA 1500,1100,500,100,-100,-1000,-2200,-2500,-1600,-400,1800,2800
230 IF C(I)<0 THEN 280
240 V=V+C(I)
250 PRINT @4: USING 260:I-1,I,C(I),V
260 IMAGE 5X,2D,X,2D,5X,5D,24X,5D
270 GO TO 320
280 F(I)=C(I)*0.9
290 V=V+F(I)
300 PRINT @4: USING 310:I-1,I,C(I),F(I),V
310 IMAGE 5X,2D,X,2D,14X,5D,5X,5D,5X,5D
320 NEXT I
330 STOP
340 END
```

Chapter 7

Setting out

The location of ground marks based on data derived either from a map or plan illustrating development proposals, or by pre-computation using some mathematical model, is a common task of the engineer in charge of construction projects. Largely as a result of the increased availability of EDM instruments facilitating the rapid and accurate determination of distance, most setting out operations are now based on data derived from coordinates. Knowing the coordinates of control stations (in terms of either the national or a local control system), and the theoretical coordinates (within the same system as the control) of the marks to be emplaced, polar coordinates (bearings and distances) may be derived and used as setting out data from control stations.

A primary advantage of using coordinates is that setting out operations need not unduly hinder, or be hindered by, construction work. Control points to be used for setting out may be located adjacent to the construction site. In addition any marks which may be disturbed in the course of construction may be readily relocated from a nearby control station. This is particularly important in respect of highway construction where it is usually necessary to relocate centreline pegs at various stages in the course of construction. The availability of conveniently placed control points also increases the opportunity for providing independent check measurements to marks set out from precomputed data.

Assuming that coordinated control points are available the main task in the setting out procedure is the calculation of 'theoretical' coordinates for the marks to be emplaced. The data required for this calculation is usually derived from development or scheme plans produced at the design stage which illustrate the proposals and provide critical dimensions. Alternatively the theoretical layout may be based on data derived directly from field measurement, or from specific design criteria, or from a combination of both.

154

7.1 Horizontal alignment of curves

The design and setting out of appropriate curves to facilitate the smooth passage of vehicles from one straight alignment to another is a typical setting out problem. Providing the radius of the arc used to connect two straights catering for high design speeds is sufficiently large, or alternatively, providing the design speed is generally low, then a simple circular curve connecting the two straights will usually suffice.

7.1.1 Highway circular curves

Circular curve design depends on the following factors (Figure 7.1):

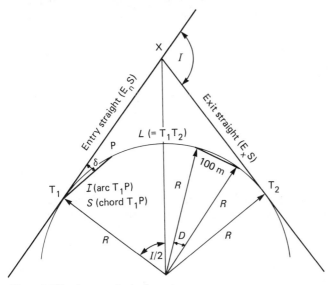

Figure 7.1 Simple curve: basic elements

1. the design velocity (V) of vehicles using the highway,
2. the configuration of the straight sections to be connected by the circular curve (i.e. the magnitude of the intersection angle, I),
3. the magnitude of the radius of the circular curve (R).

The minimum radius for a highway circular curve which may be used without transition curves (Section 7.1.4) is dependent on the adopted design speed. In the UK a range of such minimum radii is specified by the Department of Transport.

A circular curve may also be defined by its 'degree of curvature' i.e. the angle subtended at the centre of the curve by a chord or arc of standard length.

The relationship between R and the degree of curvature (D) for a standard arc length of 100 m is therefore:

$$R = \frac{100}{D^\circ (\pi/180)} \qquad (7.1)$$

The magnitude of the intersection angle is normally determined by field measurement, either by direct measurement or by deduction from the difference in the bearings of the two straights.

Given I, R, the bearing of the entry straight, the through-chainage or coordinates of the point of intersection of the straights (X), and the required spacing of centreline pegs around the curve, then the required parameters for the coordination of pegs to be set out on the centreline of the curve are as follows:

(a) *Tangent lengths*

$$T_1 X = X T_2 = R \tan (I/2) \qquad (7.2)$$

Hence the coordinate of T_1 and T_2 and the bearing and distance T_1 to T_2 are derived.

(b) *Curve length*

$$L = R I^\circ \frac{\pi}{180} \qquad (7.3)$$

Hence the required number of centreline pegs for defining the curve and the lengths of the initial and final subchords are determined.

(c) *Deflection angles*
The clockwise angle between the tangent $T_1 X$ and the line from T_1 to any point (P) on the curve is defined as a deflection angle (δ) and generally:

$$\delta'' = \frac{l}{2R} \frac{180}{\pi} \qquad (7.4)$$

where $\delta'' =$ the deflection angle in seconds of arc, and $l =$ length of the arc from T_1 to P.

For an initial arc $T_1 P_1$ of length l_i.(Figure 7.2):

$$\delta_i = \frac{l_i}{2R} \frac{180}{\pi}$$

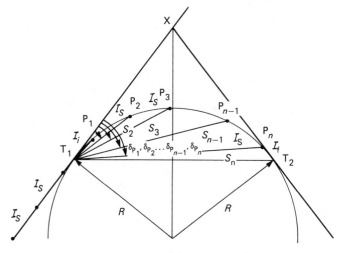

Figure 7.2 Simple curve: peg location by deflection angles and long chords

For a standard arc P_1P_2 of length l_s:

$$\delta_s = \frac{l_s}{2R}\frac{180}{\pi}$$

For a final arc P_nT_2 of length l_f:

$$\delta_f = \frac{l_f}{2R}\frac{180}{\pi}$$

The deflection angles from T_1 are therefore:

$$\delta_{P1} = \frac{l_i}{2R}\frac{180}{\pi}$$

$$\delta_{P2} = \delta_{P1} + \left(\frac{l_s}{2R}\frac{180}{\pi}\right)$$

$$\delta_{P3} = \delta_{P1} + 2\left(\frac{l_s}{2R}\frac{180}{\pi}\right)$$

$$\delta_{Pn} = \delta_{P1} + (n-1)\left(\frac{l_s}{2R}\frac{180}{\pi}\right)$$

$$\delta_{T2} = \delta_{P1} + \left(\frac{l_f}{2R}\frac{180}{\pi}\right) + (n-1)\left(\frac{l_s}{2R}\frac{180}{\pi}\right)$$

$$= \delta_{P1} + \frac{90}{\pi R}[l_f + l_s(n-1)]$$

which should equal $I/2$.

By applying the above deflection angles to the bearing of the entry straight the bearings from T_1 to all curve centreline pegs including T_2 are determined. The latter bearing should agree with that derived from Equation (7.2) providing a check on the calculation of bearings to centre line pegs.

(d) Long chords
The length (S) of the chord from T_1 to any centreline peg on the curve may be determined from the general relationship:

$$S = 2R \sin \delta \qquad\qquad (7.5)$$

Hence:

For the initial subchord $S_{P1} = 2R \sin \delta_{P1} = T_1P_1$
For P2 $S_{P2} = 2R \sin \delta_{P2} = T_1P_2$
For P3 $S_{P3} = 2R \sin \delta_{P3} = T_1P_3$
For P_n $S_{Pn} = 2R \sin \delta_{Pn} = T_1P_n$
For T_2 $S_{T2} = 2R \sin \delta_{T2} = T_1T_2$

The value for the chord T_1T_2 should obviously agree with that derived from Equations (7.2).

The theoretical coordinates of the curve centreline pegs may now be determined using their polar coordinates and the known coordinates of T_1.

Placing data in respect of each of the curve pegs is derived by 'join' using the coordinates of an individual curve peg and a conveniently positioned control point.

Example 7.1
The intersection angle between the centre lines of two straight sections of a highway is 20°. The two straights are to be connected by a circular curve of radius 500 m and pegs defining the curve are to be set out at 20 m intervals of through chainage.

The through-chainage of the intersection point is 4936.217 m. Determine the data required to coordinate the pegs defining the curve.

Solution:

Tangent lengths:

$$T_1X = XT_2 = R \tan I/2 = 88.163 \text{ m}$$

Curve length:

$$L = RI\frac{\pi}{180} = 174.533\,\text{m}$$

Through chainage of T_1

$$= \text{Through chainage of } X - T_1X$$
$$= 4936.217 + 149.997 - 88.163$$
$$= 4848.054\,\text{m}$$

Through chainage of T_2:

$$= \text{Through chainage of } T_1 + L$$
$$= 4848.054 + 174.533$$
$$= 5022.587\,\text{m}$$

Through chainage of first peg:

$$P_1 = 4860\,\text{m (initial subchord 11.946)}$$

Through chainage of last peg:

$$P_g = 5020\,\text{m (final subchord 2.587)}$$

Deflection angles:

1st deflection angle $\delta_{p1} = \dfrac{l_i}{2R}\dfrac{180}{\pi} = \dfrac{11.946}{2 \times 500}\dfrac{180}{\pi} = 0°41'04''$

2nd deflection angle $\delta_{p2} = \delta_{p1} + \left(\dfrac{l_s}{2R}\dfrac{180}{\pi}\right)$

$$= 0°41'04'' + \left(\dfrac{20}{2 \times 500}\dfrac{180}{\pi}\right)$$
$$= 1°49'49''$$

3rd deflection angle $\delta_{p3} = \delta_{p2} + \left(\dfrac{l_s}{2R}\dfrac{180}{\pi}\right)$

$$= 1°49'49'' + 1°08'45''$$
$$= 2°58'35''$$

4th deflection angle $\delta_{p4} = \delta_{p3} + 1°08'45'' = 4°07'20''$

5th deflection angle $\delta_{p5} = 5°16'05''$

6th deflection angle $\delta_{p6} = 6°24'51''$

7th deflection angle $\delta_{p7} = 7°33'36''$

8th deflection angle $\delta_{p8} = 8°42'21''$

9th deflection angle $\delta_{p9} = 9°51'06''$

$$10\text{th deflection angle } \delta_{T2} = \delta_{p9} + \left(\frac{l_f}{2R} \frac{180}{\pi} \right)$$

$$= 9°51'06'' + 0°08'54''$$
$$= 10°00'00''$$

Long chords:

For $T_1 P_1 = 2R \sin \delta_{p1} = 2 \times 500 \sin 0°41'04'' = 11.946$
$\quad T_1 P_2 = 31.939$
$\quad T_1 P_3 = 51.924$
$\quad T_1 P_4 = 71.884$
$\quad T_1 P_5 = 91.815$
$\quad T_1 P_6 = 111.715$
$\quad T_1 P_7 = 131.564$
$\quad T_1 P_8 = 151.361$
$\quad T_1 P_9 = 171.098$
$\quad T_1 T_2 = 173.648$

In triangle $XT_1 T_2$:

$$\frac{XT_1}{T_1 T_2} = \frac{\sin 10°}{\sin 160°}$$

Therefore:

$$T_1 T_2 = \frac{88.163 \times \sin 160°}{\sin 10°} = 173.648 \quad \text{(check)}$$

Program 7.1 Circular arc

```
INTERSECTION ANGLE 20
TANGENT LENGTHS 88.1634903542
LENGTH OF CIRCULAR ARC 174.532925199
LENGTH OF STANDARD CHORD 20
THROUGH CHAINAGE OF INTERSECTION POINT 4936.217

THROUGH CHAINAGE OF FIRST TANGENT POINT 4848.05350965
THROUGH CHAINAGE OF SECOND TANGENT POINT 5022.58643485

LENGTH OF FIRST SUBCHORD 11.9464903543
LENGTH OF FINAL SUBCHORD 2.58643484512

                DEFLECTION ANGLES      LONG CHORDS

POINT 1          0 41  4                 11.946
POINT 2          1 49 49                 31.941
POINT 3          2 58 34                 51.923
POINT 4          4  7 20                 71.884
POINT 5          5 16  5                 91.817
POINT 6          6 24 50                111.713
POINT 7          7 33 35                131.564
POINT 8          8 42 21                151.362
POINT 9          9 51  6                171.100
POINT10          9 59 60                173.648 CHECK
```

```
100 PRINT "PROGRAM 7.1"
110 REM THIS PROGRAM DETERMINES THE SETTING OUT DATA IN RESPECT OF A
120 REM SIMPLE CURVE USING DEFLECTION ANGLES AND LONG CHORDS (EXAMPLE
130 REM 7.1)
140 INIT
150 SET DEGREES
160 DIM D(10),L(10),M(10),S(10)
170 I=20
180 R=500
190 C=20
200 L1=4936.217
210 V=R*I*(PI/180)
220 T=R*TAN(I/2)
230 PRINT @4:"INTERSECTION ANGLE ";I
240 PRINT @4:"TANGENT LENGTHS ";T
250 PRINT @4:"LENGTH OF CIRCULAR ARC ";V
260 PRINT @4:"LENGTH OF STANDARD CHORD ";C
270 PRINT @4:"THROUGH CHAINAGE OF INTERSECTION POINT ";L1
280 L2=L1-T
290 L3=L2+V
300 X=L2/C
310 L4=C-C*(X-INT(X))
320 Y=(V-L4)/C
330 Z=INT(Y)
340 L5=(Y-Z)*C
350 D(1)=L4/(2*R)*(180/PI)
360 M(1)=(D(1)-INT(D(1)))*60
370 S(1)=(M(1)-INT(M(1)))*60
380 L(1)=2*R*SIN(D(1))
390 D(10)=L5/(2*R)*(180/PI)
400 M(10)=(D(10)-INT(D(10)))*60
410 S(10)=(M(10)-INT(M(10)))*60
420 FOR N=2 TO Z+1
430 D(N)=D(N-1)+C/(2*R)*(180/PI)
440 M(N)=(D(N)-INT(D(N)))*60
450 S(N)=(M(N)-INT(M(N)))*60
460 L(N)=2*R*SIN(D(N))
470 NEXT N
480 L(10)=2*R*SIN(D(10)+D(9))
490 PRINT @4:
500 PRINT @4:"THROUGH CHAINAGE OF FIRST TANGENT POINT ";L2
510 PRINT @4:"THROUGH CHAINAGE OF SECOND TANGENT POINT ";L3
520 PRINT @4:
530 PRINT @4:
540 PRINT @4:"LENGTH OF FIRST SUBCHORD ";L4
550 PRINT @4:"LENGTH OF FINAL SUBCHORD ";L5
560 PRINT @4:
570 PRINT @4:"        ";"DEFLECTION ANGLES";"        ";"LONG CHORDS"
580 PRINT @4:
590 FOR N=1 TO Z+1
600 PRINT @4: USING 610:N,INT(D(N)),INT(M(N)),INT(S(N)),L(N)
610 IMAGE "POINT",X,D,5X,2D,X,2D,X,2D,13X,3D.3D
620 NEXT N
630 D(10)=D(10)+D(9)
640 M(10)=(D(10)-INT(D(10)))*60
650 S(10)=(M(10)-INT(M(10)))*60+0.5
660 PRINT @4: USING 670:INT(D(10)),INT(M(10)),INT(S(10)),L(10)
670 IMAGE "POINT10",5X,2D,X,2D,X,2D,13X,3D.3D,X,"CHECK"
680 STOP
690 END
```

7.1.2 Highway compound curves

A compound circular curve comprises two circular curves of different radii, the centres of the curves lying on the same side of the common tangent as illustrated in Figure 7.3. The requirement concerning minimum radius (Section 7.1.1) is equally valid for compound curves

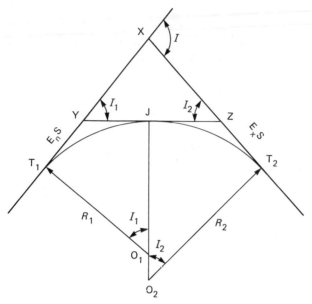

Figure 7.3 Compound curve: basic elements

and if the radii of the compound curve do not satisfy this requirement transition curves (Section 7.1.4) should be introduced connecting the compound curve to the entry and exit straights and also connecting the two curved components of the compound curve itself.

Given the through chainage or the coordinates of X, the bearing of the entry straight, the intersection angle I, the two radii (R_1 and R_2) of the compound curve, and either I_1 or I_2, then the apex distances $T_1 X$ and $T_2 X$ may be derived from the following:

$$T_1 X = \frac{R_1(1 - \cos I) + (R_2 - R_1)(1 - \cos I_2)}{\sin I} \tag{7.6a}$$

$$T_2 X = \frac{R_2(1 - \cos I) + (R_1 - R_2)(1 - \cos I_1)}{\sin I} \tag{7.6b}$$

Hence the coordinates of T_1 and T_2 may be determined.

If necessary, the tangent lengths $YT_1 (= YJ)$ and $ZT_2 (= ZJ)$ may be determined from:

$$YT_1 = YJ = R_1 \tan\left(\frac{I_1}{2}\right) \tag{7.7a}$$

$$ZT_2 = ZJ = R_2 \tan\left(\frac{I_2}{2}\right) \tag{7.7b}$$

Using the derived coordinates of say T_1 and the required spacing of centreline pegs, the coordinates of those on the arc T_1J may be determined as already described for simple circular curves, i.e. determine:

(a) the length of the arc T_1J and hence the number of centreline pegs required,
(b) the initial and final arc lengths on this section,
(c) the deflection angles and long chords from T_1,
(d) the coordinates of centreline pegs using polars from T_1.

Similarly the coordinates of the centreline pegs on the arc JT_2 may be determined from J using the common tangent YZ as the basis for deflection angles, i.e. determine:

(a) the coordinates of J by polar from T_1,
(b) the length of the arc JT_2 and hence the number of centreline pegs required,
(c) the initial and final arc lengths of this section,
(d) the deflection angles and long chords from J,
(e) the coordinates of centreline pegs using polars from J.

Example 7.2
Owing to obstructions two straights of a railway alignment are to be connected with a compound curve comprising a circular curve of 450 m radius tangential to the entry straight combined with a 550 m radius circular curve leading into the exit straight as shown in Figure 7.3. The angle of intersection of the two straights (I) is 94° and the intersection angle of the first curve (I_1) is 50°. The through chainage of X is 5160.20 m.

Determine the through chainage of T_1, J and T_2 and the deflection angles and long chords from T_1 and T_2 to J.

Solution:

$$I = 94° \quad I_1 = 50° \quad I_2 = 44°$$

Apex distances
Using formulae from Section 7.1.2:

$$T_1X = \frac{450(1 - \cos 94) + (550 - 450)(1 - \cos 44)}{\sin 94}$$
$$= \frac{481.390 + 28.066}{0.997\,564} = 510.70$$

$$T_2X = \frac{550(1 - \cos 94) + (450 - 550)(1 - \cos 50)}{\sin 94}$$

$$= \frac{588.366 - 35.721}{0.997\,564} = 553.994$$

Curve lengths:

$$T_1J = L_1 = R_1 I_1 \frac{\pi}{180} = 450 \times 50 \times \frac{\pi}{180} = 392.699$$

$$JT_2 = L_2 = R_2 I_2 \frac{\pi}{180} = 550 \times 44 \times \frac{\pi}{180} = 422.370$$

Through chainages:

Through chainage of T_1 = through chainage of $X - XT_1$
$$= 5160.20 - 510.70$$
$$= 4649.50$$
Through chainage of J   = through chainage of $T_1 + L_1$
$$= 4649.50 + 392.699$$
$$= 5042.199$$
Through chainage of T_2 = through chainage of $J + L_2$
$$= 5042.199 + 422.370$$
$$= 5464.569$$

Deflection angles:

$$\delta_{T_1J} = \frac{l_{T_1J}}{2R_1} \times \frac{180}{\pi} = \frac{392.699}{2 \times 450} \times \frac{180}{\pi} = 25°$$

$$\delta_{T_2J} = \frac{l_{T_2J}}{2R_2} \times \frac{180}{\pi} = \frac{422.370}{2 \times 550} \times \frac{180}{\pi} = 22°$$

Long chords:

$$T_1J = 2R_1 \sin \delta_{T_1J} = 2 \times 450 \sin 25° = 380.356$$
$$T_2J = 2R_2 \sin \delta_{T_2J} = 2 \times 550 \sin 22° = 412.067$$

Check on consistency:
Using T_1X as an arbitrary zero bearing:

	ΔE	ΔN
T_1X	0	510.70
XT_2	552.644	-38.645
	Σ552.644	472.055
T_1J	160.745	344.720
JT_2	391.899	127.336
	Σ552.644	472.056

Program 7.2 Compound curve

```
INTERSECTION ANGLE                      94
FIRST INTERSECTION ANGLE                50
SECOND INTERSECTION ANGLE               44
RADIUS OF FIRST CURVE                   450
RADIUS OF SECOND CURVE                  550

APEX DISTANCE; ENTRY STRAIGHT           510.700473838
APEX DISTANCE; EXIT STRAIGHT            553.994323857
LENGTH OF FIRST CURVE                   392.699081699
LENGTH OF SECOND CURVE                  422.369678983
FIRST TANGENT DISTANCE                  209.83844617
SECOND TANGENT DISTANCE                 222.214424209
CHAINAGE FIRST TANGENT POINT            4649.49952616
CHAINAGE COMMON TANGENT POINT           5042.19860786
CHAINAGE SECOND TANGENT POINT           5464.56828684
CHAINAGE FIRST INTERSECTION POINT       4859.33797233
CHAINAGE SECOND INTERSECTION POINT      5491.97989965

DEFLECTION ANGLES

FROM T1 TO J        25
FROM T2 TO J        22

LONG CHORDS

FROM T1 TO J        380.356
FROM T2 TO J        412.067


100 PRINT "PROGRAM 7.2"
110 REM THIS PROGRAM DETERMINES THE APEX DISTANCES OF A COMPOUND
120 REM CURVE AND THE SETTING OUT DATA REQUIRED IN EXAMPLE 7.2
130 INIT
140 SET DEGREES
150 READ D,D1,R1,R2,C
160 DATA 94,50,450,550,5160.2
170 I=D
180 I1=D1
190 I2=I-I1
200 A1=((R2-R1)*(1-COS(I2))+R1*(1-COS(I)))/SIN(I)
210 A2=((R1-R2)*(1-COS(I1))+R2*(1-COS(I)))/SIN(I)
220 L1=R1*I1*(PI/180)
230 L2=R2*I2*(PI/180)
240 T1=R1*TAN(I1/2)
250 T2=R2*TAN(I2/2)
260 L3=C-A1
270 L4=L3+L1
280 L5=L4+L2
290 L6=L3+T1
300 L7=C+(A2-T2)
310 D1=L1/(2*R1)*(180/PI)
320 D2=L2/(2*R2)*(180/PI)
330 S1=2*R1*SIN(D1)
340 S2=2*R2*SIN(D2)
350 PRINT @4:"INTERSECTION ANGLE ",I
360 PRINT @4:"FIRST INTERSECTION ANGLE ",I1
370 PRINT @4:"SECOND INTERSECTION ANGLE ",I2
380 PRINT @4:"RADIUS OF FIRST CURVE ",R1
390 PRINT @4:"RADIUS OF SECOND CURVE ",R2
400 PRINT @4:
410 PRINT @4:"APEX DISTANCE; ENTRY STRAIGHT ",A1
420 PRINT @4:"APEX DISTANCE; EXIT STRAIGHT ",A2
430 PRINT @4:"LENGTH OF FIRST CURVE ",L1
440 PRINT @4:"LENGTH OF SECOND CURVE ",L2
450 PRINT @4:"FIRST TANGENT DISTANCE ",T1
460 PRINT @4:"SECOND TANGENT DISTANCE ",T2
470 PRINT @4:"CHAINAGE FIRST TANGENT POINT ",L3
480 PRINT @4:"CHAINAGE COMMON TANGENT POINT ",L4
```

```
490 PRINT @4:"CHAINAGE SECOND TANGENT POINT ",L5
500 PRINT @4:"CHAINAGE FIRST INTERSECTION POINT ",L6
510 PRINT @4:"CHAINAGE SECOND INTERSECTION POINT ",L7
520 PRINT @4:
530 PRINT @4:"DEFLECTION ANGLES"
540 PRINT @4:
550 PRINT @4:"FROM T1 TO J ";"        ";D1
560 PRINT @4:"FROM T2 TO J ";"        ";D2
570 PRINT @4:
580 PRINT @4:"LONG CHORDS"
590 PRINT @4:
600 PRINT @4: USING 620:S1
610 PRINT @4: USING 630:S2
620 IMAGE "FROM T1 TO J ",6X,3D.3D
630 IMAGE "FROM T2 TO J ",6X,3D.3D
640 STOP
650 END
```

7.1.3 Highway reverse curves

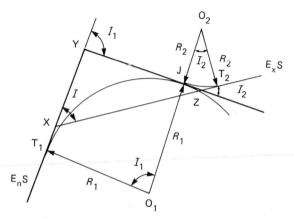

Figure 7.4 Reverse curve: $I_1 > I_2$

Reverse curves may be considered as special cases of compound curves in that they comprise two consecutive curves of the same or different radii, the curvature of one however being negative with respect to that of the other, i.e. their centres of curvature fall on opposite sides of the common tangent (Figure 7.4). The formulae in respect of compound curve apex distances are equally applicable to reverse curves. But a change in sign of some of the terms may be necessary depending on the layout and more specifically the magnitude of the intersection angles I_1 and I_2. Generally, if $I_1 > I_2$ then I is positive, and if $I_1 < I_2$ then I is negative.

For the configuration shown in Figure 7.4 ($I_1 > I_2$ and I is positive):

$$T_1X = \frac{R_1(1-\cos I)-(R_1+R_2)(1-\cos I_2)}{\sin I} \tag{7.8a}$$

$$T_2X = \frac{(R_1 + R_2)(1 - \cos I_1) - R_2(1 - \cos I)}{\sin I} \qquad (7.8b)$$

But in Figure 7.5 ($I_1 < I_2$ and I is negative):

$$T_1X = \frac{R_1(1 - \cos(-I)) - (R_1 + R_2)(1 - \cos I_2)}{\sin(-I)} \qquad (7.8c)$$

$$T_2X = \frac{(R_1 + R_2)(1 - \cos I_1) - R_2(1 - \cos(-I))}{\sin(-I)} \qquad (7.8d)$$

Having coordinated T_1, T_2, and hence J, coordinates for the centreline pegs on the entry arc (T_1J) may be determined by polar from T_1 and on the exit arc (JT_2) by polar from J or T_2 as per the compound curve calculation (Section 7.1.2).

Example 7.3 (see Figure 7.4)
Given the following data determine the apex distance T_1X and XT_2:

$$I_1 = 88° \quad I_2 = 33° \quad R_1 = 530\,\text{m} \quad R_2 = 260\,\text{m}$$

Solution:

$$I = I_1 - I_2 = 55°(+)$$

Apex distances:
Using formulae from Section 7.1.3:

$$T_1X = [530(1 - \cos(55)) - (530 + 260)(1 - \cos(33))]\operatorname{cosec}(55)$$
$$= (226.004 - 127.45)\,1.220775$$
$$= 120.312\,\text{m}$$
$$T_2X = [(530 + 260)(1 - \cos(88)) - 260(1 - \cos(55))]\operatorname{cosec}(55)$$
$$= (762.429 - 110.870)\,1.220775$$
$$= 795.407\,\text{m}$$

Example 7.4 (see Figure 7.5):
Given the following data determine the apex distances T_1X and XT_2:

$$I_1 = 59° \quad I_2 = 100° \quad R_1 = 270\,\text{m} \quad R_2 = 520\,\text{m}$$

Solution:

$$I = I_1 - I_2 = 41°(-)$$

Apex distances:

$$T_1X = [(270(1 - \cos(-41)) - (270 + 520)(1 - \cos(100))]\operatorname{cosec}(-41)$$
$$= (66.228 - 927.182) - 1.524\,253$$
$$= 1312.311\,\text{m}$$

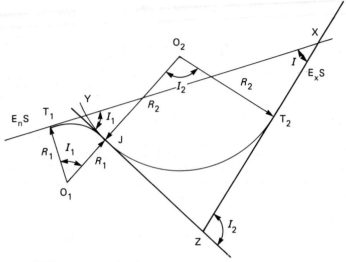

Figure 7.5 Reverse curve: $I_1 < I_2$

$$T_2X = [(270 + 520)(1 - \cos(59)) - (520(1 - c\!\slash s(-41)))] \operatorname{cosec}(-41)$$
$$= (383.120 - 127.551) - 1.524253$$
$$= -389.552 \, \text{m}$$

Program 7.3 Reverse curve

I	I1	I2	APEX DISTANCES	
			ENTRY	EXIT
55	88	33	120.313	795.407
-41	59	100	1312.311	-389.552

```
100 PRINT "PROGRAM 7.3"
110 REM THIS PROGRAM DETERMINES THE APEX DISTANCES IN RESPECT OF A
120 REM REVERSE CURVE USING DATA FROM EXAMPLES 7.3 AND 7.4
130 INIT
140 SET DEGREES
150 DIM I1(2),I2(2),I3(2),A1(2),A2(2),R1(2),R2(2)
160 PRINT @4: USING 170:
170 IMAGE 17X,"I",6X,"I1",6X,"I2",10X,"APEX DISTANCES"
180 PRINT @4:
190 PRINT @4: USING 200:
200 IMAGE 40X,"ENTRY",13X,"EXIT"
210 PRINT @4:
220 FOR J=1 TO 2
230 READ I2(J),I3(J),R1(J),R2(J)
240 DATA 88,33,530,260,59,100,270,520
250 I1(J)=I2(J)-I3(J)
260 A1(J)=(R1(J)*(1-COS(I1(J)))-(R1(J)+R2(J))*(1-COS(I3(J))))/SIN(I1(J))
270 A2(J)=((R1(J)+R2(J))*(1-COS(I2(J)))-R2(J)*(1-COS(I1(J))))/SIN(I1(J))
280 PRINT @4: USING 290:I1(J),I2(J),I3(J),A1(J),A2(J)
```

```
290 IMAGE 16X,3D,4X,3D,5X,3D,5X,4D.3D,10X,4D.3D
300 NEXT J
310 STOP
320 END
```

7.1.4 Highway transition curves

When a vehicle moving along a straight alignment is suddenly constrained to follow a circular path, the centrifugal force acting outwards from the centre of the circle through the centre of gravity of the vehicle may be such as to cause the vehicle to overturn or slide sideways away from the centre of curvature.

If the radius of the circular path is below a certain minimum it is therefore necessary to incorporate transition curves into the design.

A transition curve is a curve which provides simultaneously for (*a*) a uniform decrease of curve radius from infinity at its tangent point with an entry straight to a value *R* at its junction with the circular curve, (*b*) a constant rate of change of superelevation with respect to distance, (*c*) a constant rate of change of radial acceleration with respect to time.

Fundamental formulae in respect of these alternatives are as follows.

(*a*) In Figure 7.6:

$$r_i l_i = RL \quad \text{and} \quad \phi_i = \frac{l_i^2}{2RL} \tag{7.9}$$

where r_i = radius of curvature at any point (*i*) on the transition curve, R = radius of curvature at the junction of the transition curve and simple curve, l_i = distance along the transition curve of any point (*i*) from the point of tangency between the entry straight and the transition curve and *i*, L = length of the transition curve, and ϕ_i = the angle between the tangent to the transition curve at *i* and the entry straight.

At the end of the transition when $l = L$:

$$\phi = \frac{L}{2R} \text{ radians} \tag{7.10}$$

(*b*) In Figure 7.7:

$$\tan \alpha_i = \frac{v^2}{r_i g} \tag{7.11a}$$

and as α is usually small:

$$\alpha_i = \frac{v^2}{r_i g} \tag{7.11b}$$

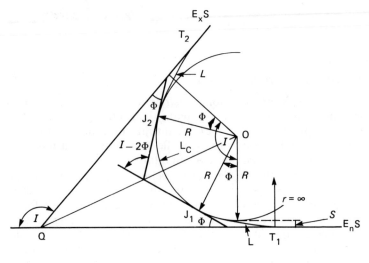

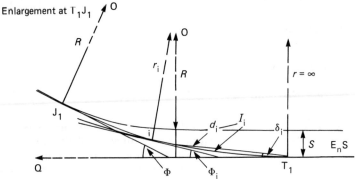

Figure 7.6 Transition curve: basic elements

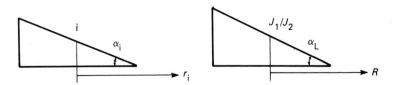

Figure 7.7 Transition curve: superelevation

where α_i = angle of superelevation at point i, v = velocity of vehicle (m/s), r_i = radius of curve at point i (m), and g = acceleration of gravity.

At the end of the transition $l = L$ and $r = R$ and expressing the velocity in km/hour (V) and R in m:

$$\tan \alpha_L = \frac{V^2}{127R} \tag{7.12}$$

(c) The length used for the transition curve must satisfy the requirement that passenger discomfort is minimized, i.e. that the change in radial force resulting from a change in curve radius is acceptable to passengers.

The following relationship can be shown to satisfy this requirement:

$$L = \frac{v^3}{ar} \tag{7.13}$$

where a = rate of change of radial acceleration.

Expressing the velocity in km/h (V) and letting $a = 0.3$ m/sec^3 (a value often adopted for major highways):

$$L = \frac{V^3}{14R} \tag{7.14}$$

Setting out data

The amount by which the circular arc connecting two straights must be moved to permit the insertion of two transition curves is termed the shift, s (Figure 7.6). It can be shown that for transition curves of small total deflection the shift bisects the transition curve and the shift itself is bisected by the transition curve. The magnitude of s is determined from:

$$s = \frac{L^2}{24R} \tag{7.15}$$

The apex distances ($T_1Q = T_2Q$) may then be deduced from:

$$T_1Q = T_2Q = (R+s)\tan\frac{I}{2} + \left(\frac{L}{2} - \frac{L^3}{240R^2} \cdots\right) \tag{7.16}$$

Theoretical positions of centreline pegs may be derived using either polar coordinates or partial rectangular coordinates based on the bearing of the entry straight and the coordinates of the first tangent point T_1.

Polar coordinates

It is convenient to determine the bearing and long chord distance to peg positions in three stages: (a) from T_1 for the entry transition curve, (b) from J_1 for the circular curve, (c) from J_2 for the exit transition curve.

(a) *Entry transition curve*
   Deflection angles
   At T_1 the deflection angle from the entry straight T_1Q to a point i distant l_i along the entry transition curve from T_1 is:

$$\delta_{Ti} = \frac{l_i^2}{6RL} \text{ radians} \qquad (7.17)$$

The deflection angles are calculated separately for each point on the curve l_s being the specified interval between centreline pegs. Through chainages are not usually maintained on the transitions. Consequently there are normally no subchords at the beginning and end of either of the two transitional sections. However, through chainages are commonly reverted to for the intervening circular arc.

Bearings from T_1 to each of the centreline peg positions are derived by applying δ_{Ti} to the known bearing of the entry straight (T_1Q).

Long chords
Chord distances (d_{Ti}) from T_1 to individual peg positions on the entry transition curve are determined from:

$$d_{Ti} = l_i - \frac{l_i^5}{90(LR)^2} + \ldots \qquad (7.18)$$

(b) *Circular curve*
   The length (L_c) of the circular curve is equal to:

$$(I - 2\phi)R\frac{\pi}{180} \qquad (7.19)$$

Knowing the through chainage of J_1 (equals the through chainage of $T_1 + L$), the length of the circular curve (L_c) and the proposed peg interval (l_s), the lengths of the initial and final subchords may be determined.

Deflection angles
The deflection angle from J_1 to a point distant l_{Ci} along the circular curve is:

$$\delta_{Ci} = \frac{l_{Ci}}{2R} \text{ radians} \qquad (7.20)$$

Bearings from J_1 to each centreline peg position are derived by applying δ_{Ci} to the bearing of the tangent at J_1 ($=$ bearing of the straight $T_1Q + \phi$).

Long chords
Chord distances (d_{Ci}) to individual peg positions on the circular curve are determined from:

$$d_{Ci} = 2R \sin \delta_{Ci} \qquad (7.21)$$

(c) *Exit transition curve*
The bearing of the tangent at J_2

$$= \text{(the bearing of the tangent at } J_1) + (I - 2\phi)R\frac{\pi}{180}$$
$$(7.22a)$$
$$= \text{(the bearing of the exit straight } QT_2) - \phi \text{ (check)}$$
$$(7.22b)$$

Deflection angles
The deflection angle from J_2 to a point distant l_i along the exit transition curve from J_2 is:

$$\delta_{Ci} - \delta_{Ti} = \frac{l_i}{2R} - \frac{l_i^2}{6RL} \text{ (radians)} \qquad (7.23)$$

Bearings from J2 to each centre line peg position are derived by applying $(\delta_{Ci} - \delta_{Ti})$ to the bearing of the tangent at J_2.

Long chords
Chord distances from J_2 to individual peg positions on the exit transition curve are equal to the corresponding distances determined from T_1.
Again the standard peg interval (l_s) is used from J_2 and the through chainages are not maintained. The through chainage of T_2 therefore equals the through chainage of $J_2 + L$.

Rectangular coordinates
The position of a peg on the transition curve (i_T) may also be defined relative to the first tangent point using the following relationships (Figure 7.8):

$$x_{iT} = l_{iT} - \frac{l_{iT}^5}{40R^2L^2} + \frac{l_{iT}^9}{3456R^4L^4} \cdots \qquad (7.24a)$$

$$y_{iT} = \frac{l_{iT}^3}{6RL} - \frac{l_{iT}^7}{336R^3L^3} + \frac{l_{iT}^{11}}{42240R^5L^5} \cdots \qquad (7.24b)$$

and at J_1 where $l = L$:

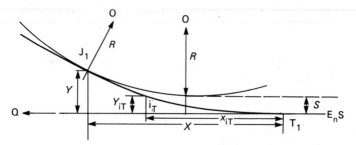

Figure 7.8 Transition curve: peg location by rectangular coordinates

$$X = L - \frac{L^3}{40R^2} + \frac{L^5}{3456R^4} \qquad (7.24c)$$

$$Y = \frac{L^2}{6R} - \frac{L^4}{336R^3} + \frac{L^6}{42\,240R^5} \cdots \qquad (7.24d)$$

For a point on the circular curve section (i_c):

$$x_{ic} = l_{ic} - \frac{l_{ic}^3}{6R^2} + \frac{l_{ic}^5}{120R^4} \cdots \qquad (7.24e)$$

$$y_{ic} = \frac{l_{ic}^2}{2R} - \frac{l_{ic}^4}{24R^3} + \frac{l_{ic}^6}{720R^5} \cdots \qquad (7.24f)$$

Example 7.5
The intersection angle (I) between two straights (Figure 7.6) is 130°
the bearing of the entry straight (T_1Q) being 271°. The straights are to
be connected by a composite curve comprising two identical transition curves connecting the straights to a central circular arc of
radius 1000 m. The design speed is 120 km/h. The through chainage
and coordinates of the point of intersection (Q) of the two straights
are as follows:

	Through chainage (m)	Coordinates (m)	
		E	N
Q	4063.54	410 542.63	240 530.66

Assuming a centreline peg interval of 20 m, determine the
theoretical coordinates of centreline pegs throughout the length of
the entry transition curve and the required superelevation at each peg
position.

Solution (see Figure 7.6):

Length of transition curve $= L = \dfrac{120^3}{14 \times 1000} = 123.429\,\text{m}$

Therefore:

$RL = 123\,429$

$\text{Shift} = s = \dfrac{123.429^2}{24 \times 1000} = 0.635\,\text{m}$

Apex distance $= T_1 Q = Q T_2$

$$= (1000 + 0.735)\tan\left(\frac{130}{2}\right) + \frac{123.429}{2}$$

$$- \frac{123.429^3}{240 \times 1000^2}$$

$$= 2207.575\,\text{m}$$

Deflection angles (at T_1) $\left(k = \dfrac{180}{\pi}\dfrac{1}{6 \times 123\,429} = 7.736\,671\,7 \times 10^{-5} \right)$

i_{20}	$= k \times 20^2$	$= 0°01'51''$	Bearing 271°01'51''
i_{40}	$= k \times 40^2$	$= 0\ 07\ 26$	271 07 26
i_{60}	$= k \times 60^2$	$= 0\ 16\ 43$	271 16 43
i_{80}	$= k \times 80^2$	$= 0\ 29\ 43$	271 29 43
i_{100}	$= k \times 100^2$	$= 0\ 46\ 25$	271 46 25
i_{120}	$= k \times 120^2$	$= 1\ 06\ 51$	272 06 51
$J_1 = i_{123.429}$	$= k \times 123.429^2$	$= 1\ 10\ 43$	272 10 43

$\phi = \dfrac{123.429}{2 \times 1000}\dfrac{180}{\pi} = 3°32'10''$

$\dfrac{\phi}{3} = 1°10'43'' = i_{123.429}$ (check)

Long chords (from T_1) $\left(k = \dfrac{1}{90 \times 123\,429^2} = 7.293\,283 \times 10^{-13} \right)$

d_{Ti20}	$= 20 - (k \times 20^5)$	$= 20.000\,\text{m}$
d_{Ti40}	$= 40 - (k \times 40^5)$	$= 40.000\,\text{m}$
d_{Ti60}	$= 60 - (k \times 60^5)$	$= 59.999\,\text{m}$
d_{Ti80}	$= 80 - (k \times 80^5)$	$= 79.998\,\text{m}$
d_{Ti100}	$= 100 - (k \times 100^5)$	$= 99.993\,\text{m}$
d_{Ti120}	$= 120 - (k \times 120^5)$	$= 119.982\,\text{m}$
$d_{TJ} = D_{Ti123.429}$	$= 123.429 - (k \times 123.429^5)$	$= 123.408\,\text{m}$

Theoretical coordinates:
For T_1:

Bearing $QT_1 = 91°$
Apex distance $= 2207.575$ m $= QT_1$

	ΔE	(m)	ΔN		E	(m)	N
Q				Q	410 542.63		240 530.66
91 00 00							
2207.575	2207.239		-38.527				
T_1				T_1	412 749.869		240 492.133
271 01 51							
20.000	-19.997		0.360				
i_{20}				i_{20}	412 729.872		240 492.493
T_1							
271 07 26							
40.000	-39.992		0.785				
i_{40}				i_{40}	412 709.877		240 492.918
T_1							
271 16 43							
59.999	-59.984		1.339				
i_{60}				i_{60}	412 689.885		240 493.472
T_1							
271 29 43							
79.998	-79.971		2.088				
i_{80}				i_{80}	412 669.898		240 494.221
T_1							
271 46 25							
99.993	-99.945		3.095				
i_{100}				i_{100}	412 649.924		240 495.228
T_1							
272 06 51							
119.982	-119.900		4.426				
i_{120}				i_{120}	412 629.969		440 496.559
T_1							
272 10 43							
123.408	-123.319		4.691				
$i_{123.429}$			$(J_1=)i_{123.429}$		412 626.550		240 496.824

Check using rectangular coordinates (see Figure 7.8):
For i_{20}

$$x = 20 - \frac{20^5}{40 \times 1000^2 \times 123.429^2} = 20.000$$

$$y = \frac{20^3}{6 \times 1000 \times 123.429} - \frac{20^7}{336 \times 1000^3 \times 123.429^3} = 0.011$$

Likewise for the remaining points:

	$-x$	$+y$
i_{40}	40.000	0.086
i_{60}	59.999	0.292
i_{80}	79.995	0.691
i_{100}	99.984	1.350
i_{120}	119.959	2.333
$i_{123.429}\ (=J_1)$	123.382	2.538

Transforming these values into ΔE and ΔN with T_1 as origin (using formulae in Section 3.3.3) a rotation (θ) of $-1°$ of the xy system is required:

$$E_{i20} = -20\cos(-1) - (0.011\sin(-1)) + 412\,749.869$$
$$= 412\,729.872$$
$$N_{i20} = -20\sin(-1) + (0.011\cos(-1)) + 240\,492.133$$
$$= 240\,492.493$$

The remaining points may be checked in a similar manner.

Superelevation:
Referring to Section 7.1.4 (*a*) and (*b*):

$$LR = 123.429 \times 1000 = 123\,429$$
$$r_i = \frac{LR}{l_i} = \frac{123\,429}{l_i}$$
$$v = 120 \times \frac{1000}{3600} = 33.333 \text{ m/s}$$
$$g = 9.806\,65 \text{ m/s}^2$$
$$\frac{v^2}{r_ig} = \frac{33.333^2}{123\,429 \times 9.806\,65} \times l_i = 0.000\,918 l_i$$
$$\alpha_i = \tan^{-1}(0.000\,918 \times l_i)$$

Hence

α_{20}	$= \tan^{-1}(0.000\,918 \times 20)$	$= 01°03'07''$
α_{40}	$= \tan^{-1}(0.000\,918 \times 40)$	$= 02\ 06\ 11$
α_{60}	$= \tan^{-1}(0.000\,918 \times 60)$	$= 03\ 09\ 10$
α_{80}	$= \tan^{-1}(0.000\,918 \times 80)$	$= 04\ 12\ 01$
α_{100}	$= \tan^{-1}(0.000\,918 \times 100)$	$= 05\ 14\ 42$
α_{120}	$= \tan^{-1}(0.000\,918 \times 120)$	$= 06\ 17\ 11$
$\alpha_{123.429}$	$= \tan^{-1}(0.000\,918 \times 123.429)$	$= 06\ 27\ 52$

Program 7.4 Transition curve and superelevation

```
DESIGN VELOCITY 120    MINIMUM RADIUS 1000

INTERSECTION ANGLE 130    THROUGH-CHAINAGE OF Q 4063.54


SHIFT 0.635    APEX DISTANCE 2207.575


                 DEFLECTION ANGLES    LONG CHORDS

POINT 1             0  1 51          20.000
POINT 2             0  7 26          40.000
POINT 3             0 16 43          59.999
POINT 4             0 29 43          79.998
POINT 5             0 46 25          99.993
POINT 6             1  6 51         119.982
POINT 7             1 10 43         123.408   (CHECK)


COORDINATES

                 BEARING       EASTINGS      NORTHINGS

POINT 1      271  1 51       412729.87      240492.49
POINT 2      271  7 26       412709.88      240492.92
POINT 3      271 16 43       412689.88      240493.47
POINT 4      271 29 43       412669.90      240494.22
POINT 5      271 46 25       412649.92      240495.23
POINT 6      272  6 51       412629.97      240496.56
POINT 7      272 10 43       412626.55      240496.82


SUPERELEVATION

POINT 1          1  3  6
POINT 2          2  6 10
POINT 3          3  9  9
POINT 4          4 12  0
POINT 5          5 14 41
POINT 6          6 17 10
POINT 7          6 27 51



100 PRINT "PROGRAM 7.4"
110 REM THIS PROGRAM DETERMINES THE COORDINATES OF THE CENTRE-LINE PEGS
120 REM ON A TRANSITION CURVE AND THE SUPERELEVATION REQUIRED AT EACH
130 REM PEG (EXAMPLE 7.5)
140 INIT
150 SET DEGREES
160 C1=4063.54
170 I=130
180 V=120
190 R=1000
200 L1=V^3/(14*R)
210 K1=R*L1
220 K2=180/PI*(1/(6*K1))
230 K3=1/(90*K1^2)
240 U=L1^2/(24*R)
250 P=INT(L1/20)
260 T=(R+U)*TAN(I/2)+L1/2-L1^3/(240*R^2)
270 PRINT @4: USING 280:V,R
280 IMAGE "DESIGN VELOCITY",X,3D,4X,"MINIMUM RADIUS",X,4D,X
290 PRINT @4:
300 PRINT @4: USING 310:I,C1
310 IMAGE"INTERSECTION ANGLE",X,3D,4X,"THROUGH-CHAINAGE OF Q",X,4D.2D
```

```
320 PRINT @4:
330 PRINT @4:
340 PRINT @4: USING 350:U,T
350 IMAGE "SHIFT",X,D.3D,4X,"APEX DISTANCE",X,4D.3D
360 PRINT @4:
370 PRINT @4:
380 PRINT @4: USING 390:
390 IMAGE 10X,"DEFLECTION ANGLES",5X,"LONG CHORDS"
400 PRINT @4:
410 DIM D(P+1),M(P+1),S(P+1),H(P+1)
420 FOR J=1 TO P
430 D(J)=K2*(J*20)^2
440 M(J)=(D(J)-INT(D(J)))*60
450 S(J)=(M(J)-INT(M(J)))*60+0.5
460 H(J)=J*20-K3*(J*20)^5
470 PRINT @4: USING 480:J,INT(D(J)),INT(M(J)),INT(S(J)),H(J)
480 IMAGE "POINT",2D,8X,2D,2(X,2D),10X,3D.3D
490 NEXT J
500 D(P+1)=K2*L1^2
510 M(P+1)=(D(P+1)-INT(D(P+1)))*60
520 S(P+1)=(M(P+1)-INT(M(P+1)))*60+0.5
530 H(P+1)=L1-K3*L1^5
540 PRINT @4: USING 550:P+1,INT(D(P+1)),INT(M(P+1)),INT(S(P+1)),H(P+1)
550 IMAGE "POINT",2D,8X,2D,2(X,2D),10X,3D.3D,2X,"(CHECK)"
560 PRINT @4:
570 PRINT @4:
580 DIM E(P+1),N(P+1),B(P+1)
590 READ X1,Y1,A1
600 DATA 410542.63,240530.66,271
610 A2=A1-180
620 X2=X1-T*SIN(A1)
630 Y2=Y1-T*COS(A1)
640 PRINT @4:"COORDINATES"
650 PRINT @4:
660 PRINT @4: USING 670:
670 IMAGE 11X,"BEARING",7X,"EASTINGS",7X,"NORTHINGS"
680 PRINT @4:
690 FOR J=1 TO P
700 B(J)=A1+D(J)
710 D(J)=INT(B(J))
720 M(J)=(B(J)-D(J))*60
730 S(J)=(M(J)-INT(M(J)))*60+0.5
740 E(J)=X2+H(J)*SIN(B(J))
750 N(J)=Y2+H(J)*COS(B(J))
760 PRINT @4: USING 770:J,D(J),INT(M(J)),INT(S(J)),E(J),N(J)
770 IMAGE "POINT",X,D,3X,3D,X,2D,X,2D,6X,6D.2D,6X,6D.2D
780 NEXT J
790 B(P+1)=A1+K2*L1^2
800 D(P+1)=INT(B(P+1))
810 M(P+1)=(B(P+1)-D(P+1))*60
820 S(P+1)=(M(P+1)-INT(M(P+1)))*60+0.5
830 E(P+1)=X2+H(P+1)*SIN(B(P+1))
840 N(P+1)=Y2+H(P+1)*COS(B(P+1))
850 PRINT @4: USING 770:P+1,D(P+1),INT(M(P+1)),INT(S(P+1)),E(P+1),N(P+1)
860 PRINT @4:
870 PRINT @4:
880 PRINT @4:"SUPERELEVATION"
890 PRINT @4:
900 V=V*(1000/3600)
910 G=9.80665
920 K4=V^2/(K1*G)
930 DIM A(P+1),Q(P+1),W(P+1)
940 FOR J=1 TO P
950 A(J)=ATN(K4*J*20)
960 Q(J)=(A(J)-INT(A(J)))*60
970 W(J)=(Q(J)-INT(Q(J)))*60+0.5
980 PRINT @4: USING 990:J,D,6X,2D,X,2D,X,2D
990 IMAGE "POINT",X,D,6X,2D,X,2D,X,2D
1000 NEXT J
1010 A(P+1)=ATN(K4*L1)
1020 Q(P+1)=(A(P+1)-INT(A(P+1)))*60
1030 W(P+1)=(Q(P+1)-INT(Q(P+1)))*60+0.5
```

```
1040 PRINT @4: USING 1050:P+1,INT(A(P+1)),INT(Q(P+1)),INT(W(P+1))
1050 IMAGE "POINT",X,D,6X,2D,X,2D,X,2D
1060 STOP
1070 END
```

Example 7.6
Using the data from Example 7.5, determine the theoretical coordinates of the first two and final two centreline pegs on the circular curve section assuming that through chainages are reverted to as from J_1.

Solution (Figure 7.9):

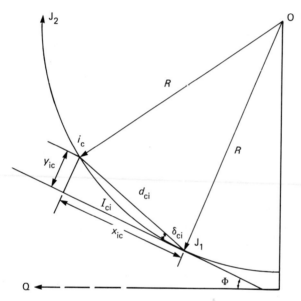

Figure 7.9 Transition curve: peg location on circular arc

$$L_T = 123.429 \text{ m} \quad L_C = (130° - 2(3°32'10'')) \times 1000 \times \frac{\pi}{180}$$
$$= 2145.494$$

Through chainage of Q = 4063.54
Apex distance $QT_1 = 2207.575$
Through chainage of $T_1 = 1855.965$
Through chainage of $J_1 = 1855.965 + 123.429 = 1979.394$
Through chainage of $J_2 = 1979.394 + 2145.494 = 4124.888$
$l_{c1} = 1980 - 1979.394 = 0.606$

$$l_{c2} = 20.606$$

.

.

.

$$l_{c107} = 2120.606$$
$$l_{c108} = 2140.606$$
$$l_{J2} = 2145.494 = L_C$$

Deflection angles (at J_1):

$$\delta_{c1} = \frac{0.606}{2 \times 100} \frac{180}{\pi} = 0°01'02.5''$$

For:

$$s = 20\,\text{m} \quad \delta_s = 0°34'23''$$

and:

$$\delta_{c2} = \delta_{c1} + \delta_s = 0°35'25''$$
$$\delta_{c107} = \delta_{c1} + 106\delta_s = 60°45'03''$$
$$\delta_{c108} = \delta_{c1} + 107\delta_s = 61°19'26''$$

$$\delta_{J2} = \delta_{c108} + \left(\frac{2145.494 - 2140.606}{2000}\right)\frac{180}{\pi}$$

$$= 61°27'50'' = \tfrac{1}{2}(I - 2)$$

Long chords:

$$d_{c1} = 2 \times 1000 \sin 0°01'02.5'' = 0.606$$

Likewise:

$$d_{c2} = 20.606$$
$$d_{c107} = 1745.007$$
$$d_{c108} = 1754.692$$
$$d_{cJ2} = 1757.032$$

Theoretical coordinates:

Bearing of tangent at $J_1 = 271° + 3°32'10''$
$$= 274°32'10''$$

	ΔE	ΔN		E	N
$J_1 - C_1$			J_1	412 626.550	240 496.824
274 33 13	−0.604	0.048			
0.606			C_1	412 625.946	249 496.872
$J_1 - C_2$					
275 07 35	−20.522	1.841			
20.604			C_2	412 606.028	240 498.665

$J_1 - C_{107}$
335 17 13	-729.542	1585.186			
1745.006			C_{107}	411 897.008	242 082.010

$J_1 - C_{108}$
335 51 35	-717.618	1601.235			
1754.688			C_{108}	411 908.932	242 098.059

$J_1 - J_2$
336 00 00	-714.649	1605.129			
1757.032			J_2	411 911.901	242 101.953

Program 7.5 Transition curve (data for circular arc component)

```
INTERSECTION ANGLE 130              MINIMUM RADIUS 1000
TOTAL DEFLECTION ANGLE FOR TRANSITION CURVE 3.5361111
APEX DISTANCE 2207.575               TRANSITION CURVE LENGTH 123.429
THROUGH-CHAINAGE INTERSECTION POINT 4063.54


CIRCULAR CURVE LENGTH 2145.49446477
INITIAL SUBCHORD 0.606000000007
FINAL SUBCHORD 4.88846476997
THROUGH-CHAINAGE FIRST TANGENT POINT 1855.965
                    FIRST COMMON TANGENT 1979.394
                    SECOND COMMON TANGENT 4124.88846477
NUMBER OF CENTRE-LINE PEGS REQUIRED FOR CIRCULAR CURVE 108

              DEFLECTION ANGLES FROM J1      LONG CHORDS FROM J1

POINT    1            0   1   2                 0.606
POINT    2            0  35  25                20.606
POINT    3            1   9  48                40.603

POINT  106           60  10  41              1735.148
POINT  107           60  45   3              1745.007
POINT  108           61  19  26              1754.692


COORDINATES

              BEARING        EASTINGS        NORTHINGS

POINT    1    274 33 12      412625.95       240496.87
POINT    2    275  7 35      412606.03       240498.66
POINT    3    275 41 58      412586.15       240500.85
POINT  106    334 42 51      411885.41       242065.72
POINT  107    335 17 13      411897.01       242082.01
POINT  108    335 51 36      411908.94       242098.06

100 PRINT "PROGRAM 7.5"
110 REM THIS PROGRAM DETERMINES THE THEORETICAL COORDINATES OF THE
120 REM CENTRE-LINE PEGS ON THE CIRCULAR CURVE SECTION BETWEEN TWO
130 REM IDENTICAL TRANSITION CURVES (EXAMPLE 7.6)
140 INIT
150 SET DEGREES
160 X=412626.55
170 Y=240496.82
180 A1=271
190 R=1000
200 I=130
210 L1=123.429
220 F1=3.5361111
230 L2=(I-2*F1)*R*(PI/180)
240 C1=4063.54
250 T=2207.575
260 C2=C1-T
270 C3=C2+L1
280 C4=C3+L2
290 L3=INT(C3/20)*20
300 L3=C3-L3
```

```
310 L3=20-L3
320 L4=L3+20
330 L5=L2-L3
340 L6=INT(L5/20)*20
350 L7=L5-L6
360 P=INT(L5/20)+1
370 PRINT @4:"INTERSECTION ANGLE ";I,"MINIMUM RADIUS ";R
380 PRINT @4:"TOTAL DEFLECTION ANGLE FOR TRANSITION CURVE ";F1
390 PRINT @4:"APEX DISTANCE ";T,"TRANSITION CURVE LENGTH ";L1
400 PRINT @4:"THROUGH-CHAINAGE INTERSECTION POINT ";C1
410 PRINT @4:
420 PRINT @4:
430 PRINT @4:"CIRCULAR CURVE LENGTH ";L2
440 PRINT @4:"INITIAL SUBCHORD ";L3
450 PRINT @4:"FINAL SUBCHORD ";L7
460 PRINT @4:"THROUGH-CHAINAGE FIRST TANGENT POINT ";C2
470 PRINT @4:"                 FIRST COMMON TANGENT ";C3
480 PRINT @4:"                SECOND COMMON TANGENT ";C4
490 PRINT @4:"NUMBER OF CENTRE-LINE PEGS REQUIRED FOR CIRCULAR CURVE ";P
500 PRINT @4:
510 PRINT @4:
520 DIM D(P+1),M(P+1),S(P+1),H(P+1),B(P+1),E(P+1),N(P+1)
530 D(1)=L3/(2*R)*(180/PI)
540 M(1)=(D(1)-INT(D(1)))*60
550 S(1)=(M(1)-INT(M(1)))*60+0.5
560 H(1)=2*R*SIN(D(1))
570 F2=20/(2*R)*(180/PI)
580 FOR J=2 TO P
590 D(J)=D(1)+(J-1)*F2
600 M(J)=(D(J)-INT(D(J)))*60
610 S(J)=(M(J)-INT(M(J)))*60+0.5
620 H(J)=2*R*SIN(D(J))
630 NEXT J
640 PRINT @4: USING 650:
650 IMAGE 10X,"DEFLECTION ANGLES FROM J1",5X,"LONG CHORDS FROM J1"
660 PRINT @4:
670 PRINT @4: USING 680:INT(D(1)),INT(M(1)),INT(S(1)),H(1)
680 IMAGE "POINT",3X,"1",8X,2D,X,2D,X,2D,18X,3D.3D
690 FOR J=2 TO 3
700 PRINT @4: USING 710:J,INT(D(J)),INT(M(J)),INT(S(J)),H(J)
710 IMAGE "POINT",4D,8X,2D,X,2D,X,2D,18X,3D.3D
720 NEXT J
730 PRINT @4:
740 FOR J=P-2 TO P
750 PRINT @4: USING 760:J,INT(D(J)),INT(M(J)),INT(S(J)),H(J)
760 IMAGE "POINT",4D,8X,2D,X,2D,X,2D,17X,4D.3D
770 NEXT J
780 PRINT @4:
790 PRINT @4:
800 PRINT @4:"COORDINATES"
810 PRINT @4:
820 PRINT @4: USING 830:
830 IMAGE 13X,"BEARING",7X,"EASTINGS",7X,"NORTHINGS"
840 PRINT @4:
850 A2=A1+F1
860 FOR J=1 TO P
870 B(J)=A2+D(J)
880 D(J)=INT(B(J))
890 M(J)=(B(J)-D(J))*60
900 S(J)=(M(J)-INT(M(J)))*60+0.5
910 E(J)=X+H(J)*SIN(B(J))
920 N(J)=Y+H(J)*COS(B(J))
930 NEXT J
940 FOR I=1 TO 3
950 PRINT @4: USING 960:I,D(I),INT(M(I)),INT(S(I)),E(I),N(I)
960 IMAGE "POINT",X,3D,3X,3D,X,2D,X,2D,6X,6D.2D,6X,6D.2D
970 NEXT I
980 FOR I=P-2 TO P
990 PRINT @4: USING 960:I,D(I),INT(M(I)),INT(S(I)),E(I),N(I)
1000 NEXT I
1010 STOP
1020 END
```

7.2 Vertical alignment

Vertical curves perform a similar function in the vertical plane as do transition curves in the horizontal plane, i.e. they smooth out changes in the gradient of two successive straights.

7.2.1 Vertical curves

A vertical curve is required wherever two gradients intersect. If the centre of curvature lies below the road formation level, it is termed a 'summit' curve. If it is above the road formation level, it is termed a 'sag' curve (Figure 7.10).

Generally, the vertical curve employed is the simple parabola, the chief geometrical property of which is that the rate of change of gradient is constant.

The simple parabola is determined by specifying the gradients of the two straights (g_1 and g_2 in terms of percentages) and deriving a length (L) for the curve which will satisfy certain requirements.

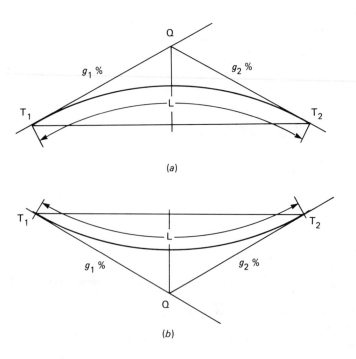

Figure 7.10 Highway vertical curve (*a*) summit curve, (*b*) sag curve

7.2.1.1 Summit curves

The requirements for single carriageway roads are governed by the need to allow for safe overtaking and the length of the vertical curve on summits is therefore determined by *sight distance*, i.e. the distance at which a vehicle driver can just see an object at the same height (h) as his eyes over the summit of a curve. The minimum overtaking sight distances recommended by the Department of Transport are based on an eye height of 1.05 m.

For dual carriageways the criterion used to determine the sight distance is the minimum stopping sight distance, i.e. the maximum distance over the curve at which a driver can see an object on the ground.

The sight distance may be assumed to be symmetrical about, and tangential to the midpoint of the curve, and also parallel to the chord T_1T_2 of the parabola (Figure 7.11).

In deducing the length of the curve two cases arise: (a) when the sight distance (S) is less than the curve length (L), and (b) when S is greater than L.

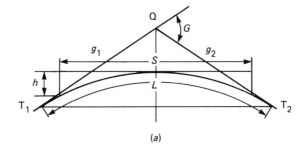

(a)

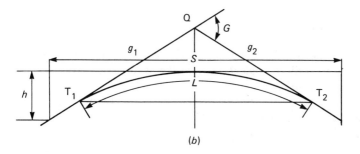

(b)

Figure 7.11 Summit curve basic elements (a) $S < L$, (b) $S > L$

(a) For $S < L$:

$$L = \frac{S^2 G}{800h} \qquad (7.26a)$$

where $G = g_1 - g_2 =$ the grade angle (the algebraic difference between the gradients adopting the convention that, in the direction of the alignment, rising gradients are positive and falling gradients are negative).

(b) For $S > L$:

$$L = 2S - \frac{800h}{G} \qquad (7.26b)$$

If, as is usually the case, the relationship between S and L is unknown, then both alternatives must be considered, one of which will produce a contradictory result and which is therefore discarded.

Sag curves

The criteria used for determining the minimum length of sag curves are alternatively (a) vehicle headlamp sight distance (= stopping sight distance), (b) motorist comfort, and (c) vertical clearance under existing structures such as bridges.

(a) *Vehicle headlamp sight distance* (Figure 7.12)

Headlamp sight distance is the main design factor for sag curves and is used to determine the length of curves as follows:

For $S < L$:

$$L = \frac{GS^2}{200(h + S\theta)} \qquad (7.27a)$$

and for $S > L$:

$$L = 2S - \frac{200(h + S\theta)}{G} \qquad (7.27b)$$

where $S\theta =$ the vertical displacement of a headlamp beam resulting from a tilt of angle θ (rads) of the headlamps.

(b) *Motorist comfort*

When considering the combined effects of gravity and radial acceleration at the low point of a sag curve the following equation is used:

$$L = \frac{GV^2}{13f} \qquad (7.28)$$

where V=design velocity in km/hour, and f=vertical acceleration at the bottom of the sag (varies between 0.15–0.6 m/sec^2 depending on the class of road)

(c) *Vertical clearance under structures* (Figure 7.13)
A minimum specified clearance between the lowest part of an existing overhead structure and the proposed road surface will automatically effect the length of the curve. To determine the length, separate equations must again be used depending on the relationship between the desired sight distance and the length of the curve, i.e. for an object on the road surface.

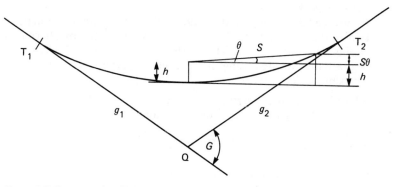

Figure 7.12 Sag curve: headlight visibility

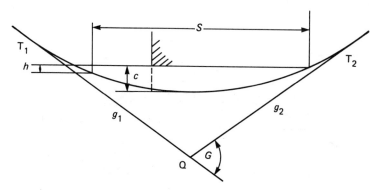

Figure 7.13 Sag curve: overhead structure

For $S < L$:

$$L = \frac{S^2 G}{800 \left(C - \dfrac{h}{2} \right)}$$ (7.29a)

and for $S > L$:

$$L = 2S - \frac{800 \left[C - \left(\dfrac{h}{2} \right) \right]}{G}$$ (7.29b)

where C = the vertical clearance between the curve and the lowest edge of the structure.

Setting out data
If the through chainage and altitude of the point of intersection (Q) is known the tangent points T_1 and T_2 may be located by alternately adding and subtracting $L/2$ to the through chainage of Q. Centreline pegs are then located using the specified peg intervals in relation to, for example, T_1.

Altitudes of tangent points (Figure 7.14)
If H_{T1} and H_{T2} are the altitudes of T_1 and T_2 respectively:

$$H_{T1} = HQ - \frac{Lg_1}{200} \quad \text{and} \quad H_{T2} = HQ + \frac{Lg_2}{200}$$ (7.30)

using the sign convention referred to earlier.

Altitudes of chainage points
The altitude of any point i on the parabola distant l_i from the first tangent point is determined from:

$$H_i = H_{T1} + \frac{g_1 l_i}{100} - \frac{G l_i^2}{200 L}$$ (7.31)

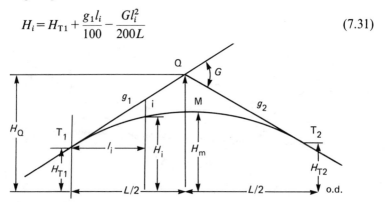

Figure 7.14 Location and heights of centre-line pegs on vertical curve

Providing the sign convention is maintained, the formula is applicable to summits and sags.

The altitude and location of the highest or lowest point on the curve (m) may be determined from:

$$H_m = H_{T1} \pm \frac{g_1^2 L}{200G} \qquad (7.32a)$$

$$l_m = \frac{g_1 L}{G} \qquad (7.32b)$$

Example 7.7

On a new road alignment, a rising gradient of 3.0 per cent meets a falling gradient of 1.5 per cent. The through chainage and altitude of the intersection point are 1506.72 m and 110.11 m respectively. The two gradients are to be connected by a parabolic curve providing for a minimum sight distance of 450 m for a driver's eye height of 1.05 m.

Tabulate the through chainages and altitudes of the tangent points and centreline pegs spaced at even 20 m intervals from the initial tangent point. Also determine the through chainage and altitude of the highest point on the curve.

Solution:

$g_1 = 3\%$ $g_2 = 1.5\%$ $\therefore$ $G = [3 - (-1.5)] = 4.5$
$S = 450$ m
$h = 1.05$ m

For $S > L$:

$$L = (2 \times 450) - \left(\frac{800 \times 1.05}{4.5}\right) = 713.333 \text{ m}$$

But $S = 450$ m therefore this solution is contradictory.

For $S < L$:

$$L = \frac{450^2 \times 4.5}{800 \times 1.05} = 1084.821 \text{ m}$$

$L/2 = 1084.821/2 = 542.411$ m
Through chainage of first tangent point (T_1)
    = 1506.72 - 542.411
    = 964.309 m
Through chainage of final tangent point (T_2)
    = 1506.72 + 542.411
    = 2049.131 m

$$\text{Altitude of } T_1 = 110.11 - \left(\frac{1084.821 \times 3}{200}\right) = 93.838\,\text{m}$$

$$\text{Altitude of } T_2 = 110.11 + \left[\frac{1084.821 \times (-1.5)}{200}\right] = 101.974\,\text{m}$$

Altitudes of centreline pegs
For P_1:

Distance from $T_1 = 20\,\text{m}$
Through chainage $= 984.309\,\text{m}$

$$\text{Altitude} = 93.838 + \frac{3 \times 20}{100} - \frac{4.5 \times 20^2}{200 \times 1084.821} = 94.430\,\text{m}$$

Likewise for remaining points:

	Distance from T_1 (m)	*Through chainage* (m)	*Altitude* (m)
P_2	40	1004.309	95.005
P_3	60	1024.309	95.563
P_4	80	1044.309	96.105
and hence to			
P_{52}	1040	2004.309	102.605
P_{53}	1060	2024.309	102.334
P_{54}	1080	2044.309	102.046
T_2	1084.821	2049.131	101.974

For the highest point:

$$H_m = 93.838 + \left(\frac{3^2 \times 1084.821}{200 \times 4.5}\right) = 104.686\,\text{m}$$

$$l_m = \frac{3 \times 1084.821}{4.5} = 723.214\,\text{m}$$

Through chainage of $M = 964.309 + 723.214 = 1687.523\,\text{m}$

Program 7.6 Vertical curve (summit)

```
CHAINAGE / ALTITUDE OF INTERSECTION POINT 110.11   1506.72
MINIMUM SIGHT DISTANCE 450
ENTRY / EXIT GRADIENT 3    -1.5
REQUIRED PEG SPACING 20

LENGTH OF CURVE 1084.82142857
```

CHAINAGE	Y	H	HT ON TANGENT	HT ON CURVE
964.309		0.000	93.838	93.838
984.309	0.008	0.600	94.438	94.429
1004.309	0.033	1.200	95.038	95.004
1024.309	0.075	1.800	95.638	95.563
1044.309	0.133	2.400	96.238	96.105

2004.309	22.433	31.200	125.038	102.604
2024.309	23.304	31.800	125.638	102.333
2044.309	24.192	32.400	126.238	102.046
2049.131	24.408	32.545	126.382	101.974

HIGHEST/LOWEST POINT

CHAINAGE 1687.524   ALTITUDE 104.686

```
100 PRINT "PROGRAM 7.6"
110 REM THIS PROGRAM DETERMINES THE DATA FOR SETTING OUT A HIGHWAY
120 REM VERTICAL CURVE GIVEN THE REQUIRED MINIMUM SIGHT DISTANCE
130 REM (EXAMPLE 7.7)
140 INIT
150 N=60
160 DIM Y(N+1),X(N+1),H(N+1),C(N+1),A(N+1),E(N+1)
170 READ G1,G2,E1,C1,D,S
180 DATA 3,-1.5,110.11,1506.72,20,450
190 G=ABS(G1-G2)
200 L=S^2/1.05*G/800
210 IF ABS(L)<S THEN 230
220 GO TO 240
230 L=2*S-800*1.05/G
240 PRINT @4:"CHAINAGE / ALTITUDE OF INTERSECTION POINT ";E1;"   ";C1
250 PRINT @4:"MINIMUM SIGHT DISTANCE ";S
260 PRINT @4:"ENTRY / EXIT GRADIENT ";G1;"   ";G2
270 PRINT @4:"REQUIRED PEG SPACING ";D
280 PRINT @4:
290 PRINT @4:"LENGTH OF CURVE ";L
300 PRINT @4:
310 PRINT @4:
320 PRINT @4: USING 330:"CHAINAGE","Y","H","HT ON TANGENT","HT ON CURVE"
330 IMAGE 4(15A),12A
340 C(1)=C1-L/2
350 H(1)=0
360 E(1)=E1-G1*L/200
370 A(1)=E(1)
380 PRINT @4:
390 PRINT @4: USING 400:C(1),H(1),A(1),E(1)
400 IMAGE 4D.3D,17X,4D.3D,12X,4D.3D,8X,4D.3D
410 K=G/(L*200)
420 X(1)=0
430 Y(1)=0
440 C(2)=C(1)+D
450 N=INT((L-(C(2)-C(1)))/D)+2
460 X(2)=C(2)-C(1)
470 Y(2)=K*X(2)^2
480 H(2)=G1*X(2)/100
490 A(2)=E(1)+H(2)
500 E(2)=A(2)-Y(2)
510 PRINT @4: USING 520:C(2),Y(2),H(2),A(2),E(2)
520 IMAGE 4D.3D,2X,4D.3D,7X,4D.3D,12X,4D.3D,8X,4D.3D
530 FOR I=3 TO N
540 X(I)=X(I-1)+D
550 Y(I)=K*X(I)^2
560 H(I)=G1*X(I)/100
570 A(I)=E(1)+H(I)
580 E(I)=A(I)-Y(I)
590 C(I)=C(I-1)+D
600 NEXT I
610 FOR I=3 TO 5
620 PRINT @4: USING 630:C(I),Y(I),H(I),A(I),E(I)
630 IMAGE 4D.3D,2X,4D.3D,7X,4D.3D,12X,4D.3D,8X,4D.3D
640 NEXT I
650 PRINT @4:
660 FOR I=N-2 TO N
670 PRINT @4: USING 630:C(I),Y(I),H(I),A(I),E(I)
680 NEXT I
690 C(N+1)=C1+L/2
700 H(N+1)=G1*L/100
710 A(N+1)=A(1)+H(N+1)
```

```
720 X(N+1)=C(N+1)-C(1)
730 Y(N+1)=K*X(N+1)^2
740 E(N+1)=A(N+1)-Y(N+1)
750 PRINT @4: USING 630:C(N+1),Y(N+1),H(N+1),A(N+1),E(N+1)
760 M=C(1)+ABS(L/G)*ABS(G1)
770 P=E(1)+G1^2*L/(200*G)
780 PRINT @4:
790 PRINT @4:
800 PRINT @4:"HIGHEST/LOWEST POINT"
810 PRINT @4:
820 PRINT @4: USING 830:M,P
830 IMAGE "CHAINAGE",2X,4D.3D,5X,"ALTITUDE",2X,3D.3D
840 STOP
850 END
```

Example 7.8

A vertical curve is required to connect a -4 per cent grade to a $+5$ per cent grade on a proposed highway. Determine the minimum length of the curve under the following separate circumstances: (a) if the design speed of the highway is 100 km/h and a maximum radial acceleration of 0.3 m/sec^2 is specified, (b) if a minimum clearance between an existing overhead structure and the road surface of 5 m and a minimum sight distance to objects on the road surface of 300 m (eye height 1.05 m) are specified.

Solution:

$$G = -4 - (+5) = 9$$

(a) $\quad L = \dfrac{100^2 \times 9}{1300 \times 0.3} = 230.78 \text{ m}$

(b) $\quad S < L, \; L = \dfrac{300^2 \times 9}{800[5-(1.05/2)]} = 226.257 \text{ m (contradictory)}$

$$S > L, \; L = (2 \times 300) - \frac{800[5-(1.05/2)]}{9} = 202.222 \text{ m}$$

Program 7.7 Vertical curve (sag)

```
ENTRY/EXIT GRADIENT -4    5

EXAMPLE 7.8(A)

DESIGN SPEED 100    MAX RADIAL ACCELERATION 0.3

MINIMUM LENGTH OF CURVE    230.769


EXAMPLE 7.8(B)

MINIMUM CLEARANCE REQUIRED 5
MINIMUM SIGHT DISTANCE REQUIRED 300
EYE/OBJECT HEIGHT 1.05

MINIMUM LENGTH OF CURVE    202.222
```

```
100 PRINT "PROGRAM 7.7"
110 REM THIS PROGRAM DETERMINES THE LENGTH OF A VERTICAL PARABOLIC
120 REM CURVE GIVEN (A) THE DESIGN SPEED OF THE HIGHWAY AND MAXIMUM
130 REM RADIAL ACCELERATION (EXAMPLE 7.8(A)) AND (B) THE MINIMUM
140 REM CLEARANCE BETWEEN AN EXISTING OVERHEAD STRUCTURE AND MINIMUM
150 REM SIGHT DISTANCE (EXAMPLE 7.8(B))
160 INIT
170 PRINT "ENTER DESIGN SPEED (KM/HR) ";
180 INPUT V
190 PRINT "ENTER MAXIMUM RADIAL ACCELERATION (M/SEC) ";
200 INPUT A
210 READ G1,G2
220 DATA -4,5
230 G=G1-G2
240 PRINT @4:"ENTRY/EXIT GRADIENT ";G1;"    ";G2
250 PRINT @4:
260 PRINT @4:
270 L1=V^2*ABS(G)/(1300*A)
280 PRINT @4:"EXAMPLE 7.8(A)"
290 PRINT @4:
300 PRINT @4:"DESIGN SPEED ";V;"    ";"MAX RADIAL ACCELERATION ";A
310 PRINT @4:
320 PRINT @4:
330 PRINT @4: USING 340:L1
340 IMAGE "MINIMUM LENGTH OF CURVE",5X,3D.3D
350 PRINT @4:
360 PRINT @4:
370 PRINT @4:
380 PRINT @4:"EXAMPLE 7.8(B)"
390 PRINT @4:
400 PRINT "ENTER MINIMUM CLEARANCE ";
410 INPUT C
420 PRINT "ENTER MINIMUM SIGHT DISTANCE ";
430 INPUT S
440 PRINT "ENTER EYE/OBJECT HEIGHT ";
450 INPUT H
460 L2=S^2*ABS(G)/(800*(C-H/2))
470 IF L2>S THEN 490
480 L2=2*S-800*(C-H/2)/ABS(G)
490 PRINT @4:"MINIMUM CLEARANCE REQUIRED ";C
500 PRINT @4:"MINIMUM SIGHT DISTANCE REQUIRED ";S
510 PRINT @4:"EYE/OBJECT HEIGHT ";H
520 PRINT @4:
530 PRINT @4:
540 PRINT @4: USING 340:L2
550 STOP
560 END
```

Bibliography

Surveying texts

Basic Metric Surveying by W. S. Whyte, Newnes-Butterworth, London, 1976

Engineering Surveying. Problems and Solutions by F. A. Shepherd, Arnold, London, 1983

Engineering Surveying Vols 1 and 2 by W. Schofield, Newnes-Butterworth, London, 1972

Land Surveying by R. J. P. Wilson, Macdonald and Evans, London, 1971

Practical Field Surveying and Computations by A. L. Allen, J. R. Hollwey and J. H. B. Maynes, Heineman, London, 1968

Principles of Surveying Vols 1 and 2 (4th edition) by J. G. Olliver and J. Clendinning, Van Nostrand Reinhold, Wokingham, 1978

Revision Notes and Plane Surveying by W. S. Whyte, Newnes-Butterworth, London, 1975

Setting Out. A Guide for Site Engineers by S. G. Brighty, Granada, London, 1975

Site Surveying. Level 3 by P. Neal, Longman, Harlow, 1985

Surveying by A. Bannister and S. Raymond, Pitman, London, 1979

Surveying for Engineers by J. Uren and W. F. Price, Macmillan, London, 1978

Computer programming texts

Basic BASIC. An Introduction to programming by D. M. Monro, Arnold, London, 1978

BASIC Programs for Land Surveying by P. Milne, Spon, London, 1984

Interactive Computing with BASIC. A First Course by D. M. Monro, Arnold, London, 1974

Index